SMITHSONIAN
INSTITUTION

UNITED STATES

NATIONAL MUSEUM

BULLETIN 245

WASHINGTON, D.C.

1965

Cincinnati
Locomotive Builders
1845–1868

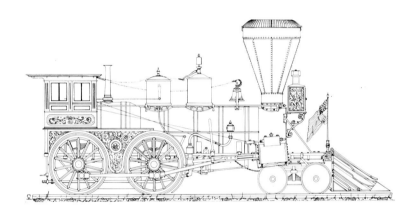

JOHN H. WHITE

Associate Curator of Transportation

MUSEUM OF HISTORY AND TECHNOLOGY

SMITHSONIAN INSTITUTION

WASHINGTON · 1965

Publications of the United States National Museum

The scholarly and scientific publications of the United States National Museum include two series, *Proceedings of the United States National Museum* and *United States National Museum Bulletin*.

In these series are published original articles and monographs dealing with the collections and work of its constituent museums—the Museum of Natural History and the Museum of History and Technology— setting forth newly acquired facts in the fields of Anthropology, Biology, History, Geology, and Technology. Copies of each publication are distributed to libraries, to cultural and scientific organizations, and to specialists and others interested in the various subjects.

The *Proceedings*, begun in 1878, are intended for the publication, in separate form, of shorter papers from the Museum of Natural History. These are gathered in volumes, octavo in size, with the publication date of each paper recorded in the table of contents of the volume.

In the *Bulletin* series, the first of which was issued in 1875, appear longer, separate publications consisting of monographs (occasionally in several parts) and volumes in which are collected works on related subjects. *Bulletins* are either octavo or quarto in size, depending on the needs of the presentation. Since 1902 papers relating to the botanical collections of the Museum have been published in the *Bulletin* series under the sub-series *Contributions from the United States National Herbarium*. Since 1959, in *Bulletins* titled "Contributions from the Museum of History and Technology," have been gathered shorter papers relating to the collections and research programs of that museum.

This work forms number 245 of the *Bulletin* series.

Frank A. Taylor
Director, United States National Museum

For sale by the Superintendent of Documents, U.S. Government Printing Office
Washington, D.C., 20402   Price $2.00

Contents

Preface

Although several detailed studies and numerous articles on eastern locomotive builders have been prepared, the early midwestern builders have been largely neglected. To many who profess an interest in locomotive history, their very existence is virtually unknown.

The history of the Cincinnati builders might well be viewed as a case study of the industry as practiced west of the Alleghenies before the Civil War. Surely, a better example would be hard to find. Except, perhaps, for McClurg, Wade & Company of Pittsburgh (a firm which built only five locomotives in the late 1830's and dropped this line of work for more profitable business), Cincinnati firms were the area's earliest commercial producers of locomotive engines. The Cincinnati shops, although largely imitators of eastern machines, introduced certain mechanical improvements. Their products, an estimated 500 engines, were to be found on lines from Panama to the western United States and were respected as more than cheap machines hastily fabricated to capture a temporary local market.

The Cincinnati Locomotive Works (Moore & Richardson) alone survived the Panic of 1857, which closed the early locomotive industry in the Midwest. The industry was not revived until the Pittsburgh (1867) and Lima (1879) shops opened after the Civil War. In Cincinnati itself, there is little question that locomotive building ceased entirely once Robert Moore closed his shop in

1868. An attempt in 1890 to establish a works for the construction of the Strong patent locomotive was unsuccessful.

Still, traces of its locomotive industry are not entirely absent in present-day Cincinnati. In December 1961 I visited Cincinnati in search of any remains of the Niles or Harkness plants. No trace of the former could be found; however, at 506 East Front Street stands an antique brick structure now occupied by the Reliance Foundry Company. While I could not say with complete assurance that the building is the original Harkness foundry, its general appearance is nearly identical to the building shown in the 1848 daguerrotype (fig. 8). I was told by an officer of the Reliance firm that the present company took over the works in about 1921 and that two foundries had previously occupied the same buildings. To the best of his knowledge the buildings had always been a foundry. Apparently, then, a foundry business has been in continuous or nearly continuous operation on this site since Anthony Harkness opened his firm in 1828.

The preparation of the history of technological subjects is frequently made difficult by the fragmentary and often contradictory information available. Rarely do the records of engineers or commercial enterprises survive. This is certainly the case with the Cincinnati locomotive industry. Although I was able to find a surprising amount of manuscript material, it was necessary to depend, in greater degree than is desirable, on newspapers and the technical press. There was little choice, however, since the companies involved have been out of business for nearly a century and their records lost or destroyed many years ago.

This work would have been impossible without the encouragement and assistance of a number of persons. My first thanks must go to Thomas Norrell, whose precise and wide knowledge of locomotive builders was invaluable in the preparation of this study. Others who generously provided information or illustrations were Harry Eddy and John McCloud of the Bureau of Railway Economics Library; Mrs. Alice P. Hook of the Historical and Philosophical Society of Ohio; Charles E. Fisher of the Railway and Locomotive Historical Society; and L. W. Sagle and M. D. Thornburg of the Baltimore and Ohio Railroad. My thanks also to the following institutions and their staffs: Cincinnati Public Library, Connecticut Historical Society, Ohio Historical Society, American Philosophical

Society, Historical Society of Pennsylvania, and the Covington Public Library. Mrs. Richard T. Keys kindly furnished the invaluable William Harkness letters, and Mr. and Mrs. Robert L. Moore contributed much useful data on Moore & Richardson. A word of thanks to my former colleague Eugene S. Ferguson for directing me to the Peale-Sellers papers. For their care and interest in preparing the manuscript my thanks to Mary E. Braunagel and Lillian Edwards. Final appreciation is due John S. Lea, Robert C. Reed, Margaret A. Pabst, Thomas C. Witherspoon, and Louise Heskett for their patient and meticulous editing, which did so much to clarify and reform this study.

The inspiration for this study ultimately rests with my parents, John and Christine White, whose high regard for their native city stimulated my interest in the industrial beginnings of Cincinnati.

John H. White
April 1965

Cincinnati
Locomotive Builders
1845–1868

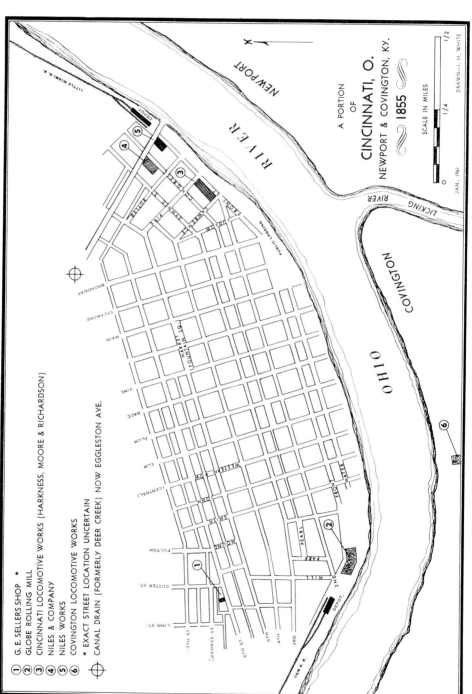

Figure 1.—Map of Cincinnati in 1855, showing location of locomotive works.

① G. E. SELLERS SHOP *
② GLOBE ROLLING MILL
③ CINCINNATI LOCOMOTIVE WORKS (HARKNESS, MOORE & RICHARDSON)
④ NILES & COMPANY
⑤ NILES WORKS
⑥ COVINGTON LOCOMOTIVE WORKS
⊕ * EXACT STREET LOCATION UNCERTAIN
CANAL DRAIN (FORMERLY DEER CREEK) NOW EGGLESTON AVE.

A PORTION
OF
CINCINNATI, O.
NEWPORT & COVINGTON, KY.
1855

SCALE IN MILES

JAN., 1961   DRAWN—J. H. WHITE

Chapter 1

Toward

The Industrial Millennium

By 1830, before any railroads were built west of the Allegheny Mountains, Cincinnati had developed into a prosperous industrial center. The city, founded in 1788, is located on the Ohio River where the Licking and the Miami and Little Miami Rivers join the Ohio. Taking full advantage of its river connections with the surrounding country, especially to the south along the Ohio River, Cincinnati had by 1810 become a center of commerce. Nine years later, with two foundries and the beginnings of what was to become a major industry—steamboat building—this city of nearly 10,000 persons was producing annually several million dollars worth of industrial products. The settlement of the surrounding country and the continued development of southern trade stimulated the commercial growth of Cincinnati to boom proportions. The city's population doubled each decade between 1820 and 1860. By the middle of the century, Cincinnati had become the largest city west of the Alleghenies and was rivaling Philadelphia's claim of having the third greatest population in the nation. It was the workshop for the South and assumed the title of Queen City, Gateway to the West. By the time of the Civil War, Cincinnati, grown to one-quarter million population, possessed a burgeoning industrial establishment which annually produced 90-million dollars worth of manufactured goods.

The industrial arts were further stimulated in 1828 by the founding of the Ohio Mechanics Institute, intended to train young men in the handling of tools, in mechanical drawing, and in other skills required by industry.

The spirit of enterprise and financial acquisitiveness was candidly portrayed by Frances Trollope in her well-known *Domestic Manners of the Americans* (1832):

> During nearly two years that I resided in Cincinnati [1828–1830], or its neighbourhood, I neither saw a beggar, nor a man of sufficient fortune to permit his ceasing his efforts to increase it; thus every bee in the hive is actively employed in search of that honey of Hybla, vulgarly called money; neither art, science, learning, nor pleasure can seduce them from its pursuit. This unity of purpose, backed by the spirit of enterprise, and joined with an acuteness and *total* absence of probability . . . may well go far toward obtaining its purpose.[1]

The reward for business activity to those of acumen in such affairs was generous, often as much as a 33⅓ percent return, but the number of failures caused by unrestricted competition and speculation severely restricted the continued enjoyment of such high profits to a very few entrepreneurs.

In an environment of thriving industrial activity, marked by busy workshops and a plentiful supply of iron, brass, and coal, as well as mechanics, it is not surprising that locomotives should be built. Nearly every large manufacturing city in this country has at some time in its history possessed concerns devoted to this industry. Quite obviously, however, in each case there had to be a local market before locomotive works could be successfully established. New England, for example, experienced this country's first railway boom, which began in the 1830's. During the next 20 years she constructed nearly 2,500 miles of railroad line, giving her nearly a third of the total trackage in the United States. This tremendous activity, virtually a mania, stimulated many machine shops to undertake locomotive construction in order to satisfy the huge local demand. Many firms were opened specifically for this purpose. The names of the New England builders—Amoskeag, Hinkley, Manchester, Lowell, Souther, Taunton, and many others— became known throughout the land. Naturally, when the New England system was completed, the eastern locomotive builders

---

[1] FRANCES TROLLOPE, *Domestic Manners of the Americans* (New York: Vintage Books, 1960), p. 43.

had to seek other markets. Many failed, while others succeeded in obtaining orders from western or southern lines by offering generous credit terms or mechanically superior products.

This same pattern of rapid expansion, repeated throughout the land as railroads were introduced, was seen in the Midwest in the late 1830's and early 1840's when many manufacturers were tempted by the prospect of huge profits in supplying machinery to the new local lines. Railroad building began in Ohio in 1835, but only 49 miles had been completed by 1840 and by 1848 only 275 miles were in operation. In 1851 the state constitution was amended so that railroad charters might be more easily obtained. This amendment, together with the expanding economy of the area, ushered in a great railway boom. The high point of activity was reached in 1854, when 587 miles of new line were opened; by 1860 Ohio possessed nearly 3,000 miles of track.[2] The effect of this rapid line construction on the establishment of a locomotive industry in Cincinnati was considerable.

But while Ohio was the first state in the old Northwest to enjoy such a concentration of railroad construction, other states in that region were not long in following her example, so that by 1860 over 9,500 miles were in operation. This encouraged the establishment of nearly a dozen western locomotive shops in such cities as Cleveland, Chicago, St. Louis, and Zanesville; however, most of these concerns were short-lived, driven out of business by the Panic of 1857.

As many readers are aware, several attempts to build locomotives west of the Alleghenies were made before the railroad era began in that section. Their lack of success amply illustrates the difficulty of establishing a locomotive-building industry before railroad construction actually began in the western regions. Therefore, it will not be amiss to discuss several of these ventures briefly in order to evaluate Cincinnati's claim to be the first city in this area to build locomotives commercially on a continuous basis.

Among the pioneer midwestern mechanics who tried their hands at locomotive construction in this all but antediluvian period of the mechanical arts were Joseph Bruen and Thomas Barlow, who in

---

[2] For a more complete history of midwestern and Ohio railroads, see B. H. MEYER, *History of Transportation in the United States Before 1860.*

Figure 2.—CINCINNATI IN THE 1850's.

1826 or 1828 built a miniature locomotive in Lexington, Kentucky. The little machine, large enough to pull a car and several passengers, was exhibited in Louisville, throughout Indiana and Ohio, and as far south as New Orleans. In 1832 Bruen built a full-size locomotive for use on the Lexington and Ohio Railroad, but it proved a failure. (A more complete description of this machine and Bruen's activities can be found in T. D. Clark's excellent article, "The Lexington and Ohio Railroad—A Pioneer Venture." [3])

Another early attempt at locomotive construction was that by Francis Shield of Cincinnati. Shield was born in England, the homeland of the locomotive, in 1783. Trained as a machinist, he settled in Cincinnati in 1819. It has not been determined when he first began experimenting with locomotives, and the subsequent history of other such machines built by him is equally elusive. He is credited with building a railway engine for a Mr. Grover of Lexington, Kentucky, before 1830. [4] Whether this had any connection with Bruen or the Lexington and Ohio Railroad must remain

[3] *Register: Kentucky State Historical Society* (January 1933), pp. 9–28.

[4] From the *Cincinnati Advertiser and Ohio Phoenix* (July 31, 1830), as reprinted in *In Memoriam*, p. 248.

in question until more conclusive evidence is forthcoming. By late May 1830 a second locomotive had been completed.[5] This model, first exhibited in Cincinnati, is described in the following newspaper article:

The Messrs. Shield of this place, who some time since invented and built for Mr. Grover, of Lexington, Kentucky, a locomotive engine, called the *Western Star*, have made some very ingenious and important improvements on that plan, which can be seen exemplified in the *Cincinnati*, a new railroad engine and car, which they have recently constructed, and which now is exhibiting at the Amphitheater in this city. We are assured that the engine is capable of overcoming an elevation of eighty or ninety feet in the mile with the greatest ease, very little difference being perceptible between the level and the above-mentioned elevation, as to speed. The principal improvements in engine consist of compactness and the application of power direct to the carriage, while the arrangement of the boiler is supposed to present a greater surface of water to the action of caloric than any other boiler heretofore constructed for a like purpose, consequently a greater quantity of steam must be generated in a given time from the same quantity of fuel.

The boiler itself is unique, and can not fail to interest the scientific and learned. The movement of the machine on the railway is admirable, and we should suppose that the engine, with the car attached carrying four persons, would perform the distance of from eight to sixteen miles per hour, at pleasure; and we have been assured by competent judges that when placed on a straight and level rail, it would perform with ease thirty miles an hour, carrying from four to eight persons. The workmanship of the whole is admirably finished, and is very creditable to Mr. Shield and the very ingenious young mechanic, his son. We understand that Mr. Shield intends very shortly to visit Baltimore and the Eastern cities with his improvement, so that the machine will remain here but a short time, and we can not but recommend an early and general attendance at the Amphitheater to witness this very fine specimen of the mechanical arts.[6]

No precise description has been found, but several facts of some assistance in picturing the engine have been culled from the numer-

---

[5] Sometime between 1830 and 1832 Shield built a full-size locomotive which was sold to the Pontchartrain Railroad in Louisiana for $1,000. The engine was a failure, but was retained for use as a stationary power plant at the railroad's repair shop.

[6] *In Memoriam*, p. 248.

ous newspaper accounts. One curious notice observed that, exclusive of the boiler, the engine disassembled could fit in a box 2 feet long, 1 foot wide, and 1 foot deep.[7] It was equipped, reports another account, with a single cylinder and a condenser.[8] The track was an oblong circle about 30 feet in diameter.[9] The model was exhibited in Rochester, Philadelphia, Baltimore, and, not unlikely, in other cities. It was sold in Baltimore by Shield in late October or early November 1830.[10]

The introduction to the industrial East of an advanced mechanism from the primitive West, a reversal of the then usual stereotype, is a significant development. Interestingly, the Shield model was shown in Philadelphia nearly a year before Matthias Baldwin built his miniature locomotive for display at the Peale Museum. Baldwin's model, however, came to the attention of the officials of a local Philadelphia railroad looking for a domestic source of motive power; this coincidence, and not solely his mechanical skill, launched him on his successful career as a locomotive builder.

Five engines built in Pittsburgh between 1834 and 1837 by McClurg, Wade & Company just about complete the record of midwestern locomotive construction prior to 1845.[11] McClurg, Wade & Company apparently did not become more fully engaged in locomotive manufacture because of the small market for such machines in the locality, and because of the availability of other more profitable business. It is also likely that their engines did not compare well with the more sophisticated products of the eastern shops.

With this as background, we will trace the affairs during the years 1845–1868 of the five locomotive concerns located in Cincinnati: Anthony Harkness & Son, the Cincinnati Locomotive Works, Niles & Company, G. E. Sellers, and the Covington Locomotive Works.

---

[7] *Poulson's American Daily Advertiser* (Philadelphia), August 12, 1830.

[8] *Daily Chronicle* (Philadelphia), August 21, 1830.

[9] Ibid., and *Baltimore Republican and Commercial Advertiser*, October 30, 1830.

[10] *In Memoriam*, p. 248.

[11] Their first engine, the *Pittsburgh*, was completed in September 1835 for the Allegheny Portage Railroad. It was copied from the *Boston*, built by the Mill Dam Foundry in 1834 (*Railway & Locomotive Historical Society Bulletin* no. 53, p. 67).

Chapter 2

Anthony Harkness and the Cincinnati Locomotive Works

The founder of the Cincinnati locomotive industry, Anthony Harkness, was born in Portsmouth, Rhode Island, on July 10, 1793.[12] He learned the machinist trade in the prosperous industrial town of Paterson, New Jersey. Part of his apprenticeship was spent with Thomas Rogers, who was to become one of the best known locomotive builders in this country. By about 1818 Harkness had risen to the superintendency of a cotton factory. Due to the financial panic of 1819, however, he migrated to Cincinnati to seek employment, believing that conditions might be better in the West, where there was a demand for skilled mechanics. His hopes were not well founded, for hard times had hit Cincinnati; discouraged, he left the city and tried his hand at farming. This proved a failure, for he had little interest in agriculture and soon tired of the isolated farm life. At about this time, the early 1820's, he met another Cincinnati mechanic, James Goodloe, with whom he formed a partnership. On the northeast corner of Broadway and Pearl Streets they opened a small machine shop and copper foundry. On the second floor of the building they also established a mill for spinning wool and cotton. The partnership ended in 1827 because they were ". . . two men so much alike that they could not agree." [13]

---

[12] This account of Harkness' life is based, unless noted otherwise, on information supplied by Mrs. Richard T. Keys, whose husband was a descendant of Harkness, and on the obituary of him in the *Cincinnati Daily Gazette*, May 12, 1858.

[13] MOORE, *Autobiographical Outlines*, p. 49.

Harkness retired from the firm with $4,000 and borrowed another $2,000. With this, in the summer of 1828, he built a new shop on the north side of Front Street, just east of Lawrence Street. This modest plant proved to be the nucleus for the giant Harkness factory which, within a few years, occupied nearly an entire block.

Not long after the new plant opened, some French sugar planters from Louisiana gave Harkness a large order for sugar mills and engines to drive them. Realizing the risk involved and feeling that his shop was ill equipped to turn out the work, Harkness at first refused to consider the proposition. However, the Bank of the United States agreed to lend him the funds necessary to complete the $30,000 order. This plunge proved to be the founding of Harkness' fortune; from this time forward he was known not only as a mechanic but also as a capitalist involved in many ventures.

In connection with the foundry business, Harkness became associated with several partners. While the exact nature and dates of these partnerships are difficult to determine, a brief outline can be reconstructed from various items found in the city directories and the *American Railroad Journal*. From about 1829 to 1840 the firm was known as the Hamilton Foundry, operated by the proprietors Pierce, Harkness, and Voorhees in various combinations, viz, Pierce, Harkness & Company (1832)[14] and Harkness, Voorhees & Company (1837).[15] By the early 1840's Harkness had assumed control of the business. Harkness' ventures with other investments brought him into close association with many prominent capitalists of the city. In 1844 he joined Jacob Strader and Samuel Fosdick in establishing the Franklin Cotton Mill.

It is quite likely that Harkness' association with Strader prompted him to enter the locomotive business. At the time they established the cotton mill, Strader was the treasurer of the struggling Little Miami Railroad, the first line to enter the city. Beginning at Cincinnati in 1838, construction had been pushed to Milford in late 1841, but no substantial operation was achieved until the line reached Xenia in 1845. Strader was, naturally, well aware of the financial difficulties and numerous other problems connected with

---

[14] *American Railroad Journal* (1832), vol. 2, p. 585.

[15] Ibid. (1837), vol. 6, p. 54. Robert Moore (see p. 48) spelled the name Voorhese.

the opening of the line, not the least of which was the chronic shortage of motive power. In November 1846 only two engines were in operating condition, forcing the road to turn away many shippers.[16]

The demand for new locomotives was so great that most eastern builders were refusing orders. These facts were undoubtedly brought to Harkness' attention by his associate Strader, with the obvious suggestion that locomotives be built in Harkness' shop. By this time, 1845, Harkness was deeply involved in building steamboat engines and boilers; his shop was well equipped for this type of work, so the building of locomotive engines presented no particular difficulties.

The designer of Harkness' earliest locomotives was Alexander Bonner Latta (1821–1865), who was born in Ross County, near Chillicothe, Ohio. When his family moved to Cincinnati, he apprenticed in the foundry and shipbuilding trade in that city. In 1841 he went to Washington, D.C., pursuant to a patent matter, and while there met Harkness, who had come east to secure a planing machine for iron. Young Latta impressed the older mechanic so much that he was hired to superintend the Harkness foundry. At first he succeeded brilliantly. He built a giant lathe with a $7\frac{1}{2}$-foot swing and 35-foot ways, and an equally immense planing machine which weighed an estimated 30 tons and was capable of handling work measuring 6 feet wide, 5 feet high, and 24 feet long.[17] G. E. Sellers said the planing machine was ". . . a masterpiece of mechanism . . . large enough to face, fo[u]r side pipes of the largest size steamboat cylinders then constructed, also valve faces of blowing cylinders, for iron furnaces. . . ."[18]

It is small wonder, then, that Harkness confidently assigned Latta the task of building his first locomotive. At this, however, he was not successful. Edwin Price, who as a boy in the late 1840's was employed by Latta at his Buckeye Works, recalled Latta's attempt at locomotive building:

> He built two only. They were his own Idea and Invention. He would not copy. The first was nine months in building, the second one about six. The[y] cost about twice the amount of Money Harkness received

---

[16] *Cincinnati Daily Commercial*, November 16, 1846.

[17] *Cincinnati Daily Gazette*, February 5, 1859.

[18] *American Machinist* (December 19, 1889), vol. 12, p. 2.

11

Figure 3.—A. B. LATTA, 1821–1865. (*Photo courtesy Cincinnati Chamber of Commerce.*)

for them. As this was nonprofitable a change was made and Mr Latta was succeeded by Mr. William Van Loan [Loon] who was Master Mechanic of the Little Miami R.R., whose place was filled by his Brother in law Mr C T. Ham. Van Loan built a light passenger Engine after the Roger[s] Pattern turned them out quite lively making a paying business for Mr. Harkness, diminishing the time of building from 9 & 6 Months down to three weeks. Latta being offered the Cotton Machinery department and the same pay would not except [sic] but resigned . . . .[19]

Price's recollection neatly summarizes Latta's early attempts, but the local press and the annual reports of the Little Miami Railroad also carried some valuable items on these engines. The earliest

---

[19] Edwin Price reminiscences [1829–1901] (fragmentary MS., U.S. National Museum, Washington, D.C.).

Figure 4.—LATTA'S MAMMOTH LATHE of 1848. Note the locomotive driving wheel being turned. (From *Farmer and Mechanic*, August 31, 1848.)

notice found was in the report of the Little Miami issued in December 1845, which noted that Harkness was building a 13-ton, 8-wheel, passenger engine. Construction of the first locomotive was thus begun sometime before December 1845. Eight months later the *Cincinnati Daily Gazette* of August 22, 1846, reported that Harkness had two locomotives under construction for the Little Miami Railroad. The *Cincinnati*, the first to be finished, was not placed on the road until November 15, 1846. Since one engine was started sometime before December 1845, the time spent in its construction must have been at least a year, although Price stated that the first engine built by Harkness and Latta was completed in nine months. Could it be, then, that there was an earlier engine, built before the *Cincinnati*, which proved a total failure? And, assuming the existence of an earlier, unsuccessful machine, might it have been rebuilt into the *Cincinnati*? Some support is given to this speculation by a statement in Charles Greves' *Centennial History of Cincinnati*, which says that the first locomotive built by Harkness for the Little Miami was called the *Bull of the Woods*.[20] Furthermore, Richard A. Thomas, a machinist at the Little Miami Railroad's Pendleton Shops from the 1840's on, recalled an engine of that name.[21] There is, however, no mention of an engine so named in the annual reports of the Little Miami Railroad during the period, nor could any be found in

---

[20] Vol. 1, p. 661.
[21] WATKINS, History of the Pennsylvania Railroad, vol. 2, pt. 2, p. 40.

contemporary local newspapers consulted. This lack of contemporary evidence persuades me to suggest that the *Bull of the Woods* was a total failure and was rebuilt into the *Cincinnati*. That is, the seeming contradictions present in the foregoing accounts are resolved if we assume that a 13-ton, 8-wheel passenger engine, the *Bull of the Woods*, was built between late 1845 and August 1846 (the 9-month period stated by Price); proving a failure, it was rebuilt into a 17-ton, 8-wheel engine for freight service and renamed the *Cincinnati*.

While the true history of the *Bull of the Woods* will undoubtedly remain an enigma, a fine account of the *Cincinnati* was given in the *Cincinnati Daily Commercial* for November 16, 1846.

> The new locomotive built by Mr. Anthony Harkness, of this city, for the Little-Miami Railroad Company was put upon the track on Saturday [Nov. 14], completely finished. It is the most beautiful piece of mechanism we have ever seen, and we doubt not but that it will come practically up to its appearance. The several eastern built locomotives, now on the line do not compare with this, Cincinnati's first effort, either as regards extreme beauty of finish or strength of machinery and we look upon it as an achievement which not only our mechanics of the East, but those across the water may well envy. It was built under the supervision of Mr. Latta, a most ingenious and accomplished workman, many specimens of whose skill now adorn the extensive works of Mr. Harkness. This locomotive is called the "Cincinnati", and is the largest on the line— possessing power sufficient to propel thirty or forty cars.
>
> Its weight is about seventeen tons—cost, $7,000. Mr. Latta has made some alterations from the old plans in the machinery of this locomotive, the utility of which is yet to be tested. Among them, we may mention that the steam pipe is placed outside, instead of inside the boiler. There are other alterations which, when thoroughly tested, we may notice hereafter.
>
> *　　*　　*　　*　　*
>
> The "Cincinnati" and its accompanying tank, was taken to the road from the foundry by a temporary track, made of strong timbers, laid in the street. The machines were driven along by steam power and as they passed over the different segments, those left behind were taken up and put in front, and so on it was continued until it reached its destination. When the "Iron Horse" felt his hoof upon the track which it was his destiny to tread, he gave a neigh which was triumphantly answered by the foundry boys crowded on his back.

Elsewhere in the same edition of the paper it was said of the

14

Cincinnati: "The new one, which we have noticed in another column, and which was destined for the freight business, leaks so badly in the flues, that it will take several days to put her in order."[22]

Several days later the *Commercial*, apologizing for an unintended criticism of the *Cincinnati*, commented: "In our notice of the flues of the *Cincinnati* burning out, we did not intend any reflection upon the engine. . . ." [23] This gesture to local pride not only betrays a desire not to offend home industry but, more important, substantiates Price's recollections made some fifty years later, of Latta's failings as a locomotive builder. The same report mentioned that the *Cincinnati* was expected to be ready for service in the next few days and that the Little Miami had ordered five more engines from Harkness. The Mad River and Lake Erie Railroad was also to contract for two locomotives.

While speaking of the *Cincinnati* in the annual report of the Little Miami Railroad for 1846, Jeremiah Morrow, president of the road, indicated more than a casual interest in the establishment of a locomotive works in the city.

> If a class of locomotives, suited to our purpose, can be built at home, it is unquestionably good policy to encourage the manufactory. We have had one built by Anthony Harkness, of this city, which, in neatness of finish is equal to any on our road. On trial it was found defective in parts of the works, as might be expected in the first effort to construct a complicated piece of machinery. It will require but little time and a small expense to correct the defects, when, in the opinion of those competent to judge, it will be capable of doing good work on the road. Mr. Harkness has a second engine for our road, in a state of forwardness, in the construction of which, the defects in the first will be guarded against. Should he succeed according to our expectations in furnishing a well finished and perfect engine, the cost and delay in the transportation from the East may hereafter be avoided.[24]

The Little Miami was to be one of Harkness' best customers, purchasing over 30 engines from him during the next 10 years.

Not long after Harkness had started work on his first two locomotives he wrote to the Baldwin locomotive works in Philadelphia regarding the possibilities of building in his shop the flexible-beam

[22] *Cincinnati Daily Commercial*, November 16, 1846.
[23] Ibid., November 20, 1846.
[24] *Fourth Annual Report of the Little Miami Railroad*, Cincinnati, December 1846.

15

locomotive patented by M. W. Baldwin in 1842. Designed for freight service, the flexible-beam locomotive was so constructed that the front two driving wheels could shift laterally, thus allowing a 6- or 8-couple engine to pass around sharp curves with greater ease than would be possible with an engine of the usual construction. Harkness' letter is reproduced below.

Cincinnati Dec. 24th 1845

Messrs. Baldwin & Whitney
Gent

I take the liberty to write to know from you, if I could make an arrangement with you to Build those Six Wheel freight Engines of the same constructions as those you furnished for the Little Miami Road, and what you would Charge for the privilege on each Engine. I now have two heavy Eight Wheel Engines in hand for that Road, and they will, together with the other Roads in this Section of the Country use the Eight Wheel Engines for freight, unless those freight Engines can be built here, owing to the great Expence Delay and trouble in getting them out, and Difficulty in making the Repairs to them here, without any of the Patterns being here that they were built from, if you have any Doubt about making the Arrangement, you had better write Mr. W. H. Clements the Superintendent of the Little Miami Rail Road on the subject.   your early reply to the above Communication is desired.

Yours with Respect,
Anthony Harkness [25]

Baldwin's reply, dated January 17, 1846, stated that they had no interest in selling the patent, but that a licensing arrangement might be possible.[26]   Baldwin went on to say that they had experienced no difficulty in selling the flexible-beam engines and expected them to capture a larger portion of the locomotive market in future years (the flexible-beam engine has been said to be the basis of Baldwin's fortune; several hundred were built between 1842 and about 1866). His letter continued:

We should not therefore be willing to sell the right for others to make them for a less sum than we should deem a fair profit for the manufacture and we should not be willing to sell at that price, only for a limited number, as we have it in contemplation to start an establishment for the manufacture of engines somewhere in the Ohio Valley as soon as

---

[25] M.W. Baldwin papers (MSS., Historical Society of Pennsylvania, Philadelphia).
[26] Ibid.

16

the demand for them from that quarter shall justify such an undertaking, providing a continuous railroad from this city to that valley is not previously constructed. When that is done, we can transport engines from here to roads west of the Ohio for a comparatively small sum. . . .

Baldwin ended with the request that Harkness make an offer for the right to manufacture engines after the flexible-beam patent. His letter apparently dampened Harkness' interest, for no counter offer was made. The matter was reopened a month later by a visit of William H. Clements, Superintendent of the Little Miami Railroad, to the Baldwin works. Clements was apparently convinced of the superiority of the flexible-beam locomotive for freight service and, realizing that Baldwin could not furnish sufficient numbers of these machines, was desirous of seeing a Cincinnati supplier undertake their manufacture. Baldwin wrote to Harkness again on February 7, 1846, offering to sell a license for the manufacture of from 12 to 20 locomotives at $800 each,[27] and adding that he would not make such terms to an eastern competitor. He also commented that since the $800 fee amounted to the usual transportation cost of shipping an engine west, Harkness would have an opportunity to make the same profit that an eastern builder could expect. Again Harkness showed no interest and no agreement was reached. Needless to say, Baldwin did not open a locomotive plant outside the Philadelphia area.

After Latta left Harkness in 1846 he founded the Buckeye Works with his brothers, Edmundson [28] and Findlay. Several years later, in 1852, he built, in conjunction with Abel Shawk, a steam fire engine which in error has been popularly called his invention and also cited as the first steam fire engine. He returned to locomotive design in 1856 and induced the Boston Locomotive Works to build a coal-burning locomotive after his design. This machine proved a total failure; however, the boiler was salvaged for stationary use at the builder's shop.[29] The next year Latta issued a catalog advocating several improvements for locomotives, including the articulated

---

[27] Ibid.

[28] Edmundson's name appeared in the city directories as Edminston; G.E. Sellers knew him as A.B. Latta's crippled brother, Eben.

[29] *Railroad Advocate* (December 13, 1856), vol. 3, p. 4.

Figure 5.—LATTA'S PROPOSED DUPLEX LOCOMOTIVE of 1857.
(From *American Machinist*, April 4, 1912.)

locomotive shown in figure 5.[30] The design of this machine permitted a more widespread distribution of the engine's weight as well as utilization of the dead weight of the tender for traction. The striking similarity of this plan to the ill starred "Duplex" locomotives tested by the Southern Railway before World War I should be apparent to many readers. Latta's son, Griffin T. Latta, claimed in 1929 that a locomotive of this type was built for the Boston and Maine Railroad, but no evidence has been found to verify his statement.[31]

Latta's last attempt at locomotive building was the construction of a steam dummy locomotive for use on the newly opened street railway in Cincinnati. The *Scientific Artisan* of November 26, 1859 (p. 117), noted that the Latta brothers had such an engine under construction at their Buckeye Works. It was tested early in March 1860 (fig. 6). The locomotive was powered by a 6-horsepower vertical portable engine, manufactured as a stock item by the Buckeye Works, which burned coke to reduce smoke (fig. 7). While

---

[30] The *American Machinist* (April 4, 1912, vol. 36, p. 533) discussed this catalog briefly and reproduced the engraving of the *Economist*. An earlier reference to the catalog was made by Latta in the *Railroad Advocate* (December 27, 1856), in which he defended his coal-burning locomotive and noted "I am now preparing a circular . . ." on locomotive improvements. In its issue of April 25, 1857, the *Advocate* noted the publication of the circular and reviewed Latta's idea for wrought-iron driving wheels as described therein.

[31] G.T. Latta to C.W. Mitman (curator, U.S. National Museum), October 31, 1929, data file, U.S. National Museum.

Figure 6.—Street-Railway Locomotive tested on Cincinnati horse railway in 1860. Built and designed by A. B. and E. Latta. (From *Scientific Artisan*, vol. 2, November 5, 1859.)

undoubtedly underpowered, the curious little machine was a mechanical success, pulling a car loaded with 41 passengers. Its undoing was that which largely defeated the use of dummy engines on city streets: horses became terrified when it passed. The *Cincinnati Gazette* of March 28, 1860, noted: "Scarce a horse could be coaxed or compelled to pass it and within the brief space of twenty-four hours not less than two accidents was the result."

Since we have such a fragmentary record of Latta's career, it is hardly possible to make any conclusive estimate of his mechanical abilities, but in view of the foregoing facts, I am astonished by the glowing and unqualified praise of Latta's engineering genius that has appeared since his death. G. E. Sellers spoke of him as a "progressive, skillful, and inventive mechanic,"[32] while Robert

---

[32] *American Machinist* (December 19, 1889), vol. 12, p. 2.

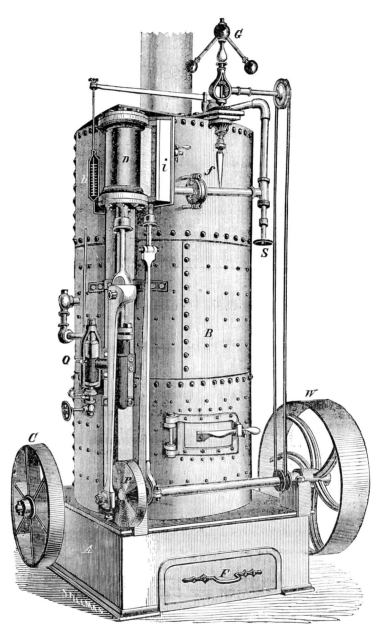

Figure 7.—THIS 6-HORSEPOWER PORTABLE ENGINE was adapted to power Latta's street-railway locomotive. (From *Scientific Artisan*, vol. 3, March 17, 1860.)

Moore, who also knew him quite well, spoke of Latta in the highest terms.[33] It might be fair to say of Latta that while he attempted to be original in all his works and was devoted to the honest perfection of mechanisms, his genius was not above the production of several dismal failures.

[33] MOORE, *Autobiographical Outlines*, pp. 56–58.

Figure 8.—A. HARKNESS & SON FOUNDRY as it appeared in 1848. It is the large 2- and 3-story building in the central foreground. (*Photo courtesy Cincinnati Public Library.*)

21

As we have already seen, Latta was succeeded by William Van Loon, who put Harkness locomotive production on a commercially successful basis. During 1847 about seven locomotives were built for the Little Miami, the Madison and Indianapolis, and the Mad River and Lake Erie railroads. The next year Harkness was so encouraged by the prospects for expanding locomotive work that he acquired the shops of Yeatman and Shield, expanding his giant Front Street plant along the entire block from Lawrence to Pike Streets.[34] The construction of cars for the Madison and Indianapolis Railroad has been reported, although perhaps erroneously. Harkness and his successors were, however, heavily engaged in the production of car wheels.[35] Locomotives were also built for stock during this early period, for the *Commercial* noted: "Mr. Harkness has now on hand five locomotives."[36]

An excellent picture of Harkness' foundry can be seen in figure 8, which is reproduced from one of a series of daguerreotypes taken by W. S. Porter in the late summer of 1848.[37] The Harkness plant is the large 2- and 3-story brick structure in the center foreground of the view. The smokestack and cupola indicate that the 2-story section of the building, on the left, was the foundry. Several railroad car wheels can be seen on the Lawrence Street side of the foundry.

Not long after the shop had been expanded, William, the only son of Anthony Harkness, was taken into the firm, which assumed the new corporate title of A. Harkness & Son. Since his father devoted himself to other investments, William was in the next few years to assume a greater part of the responsibility for the foundry's operation.

The Columbus, Ohio, newspapers of February 25, 1850, reported that the first train into that city over the newly opened Columbus and Xenia Railroad was pulled by the "brand new" *Washington*, a product of A. Harkness & Son. The *Washington* was described as weighing $19\frac{1}{4}$ tons, costing $8,000, having a maximum speed of 53

---

[34] *Cist's Advertiser* (Cincinnati), March 7, 1848.

[35] *Cincinnati Daily Commercial*, April 16, 1848.

[36] Ibid.

[37] These views are described in Carl Vitz's "The Cincinnati Water Front, 1848," *Bulletin: Historical and Philosophical Society of Ohio* (1948), vol. 6, pp. 28–39. [See plate 6 for a view of the Niles Works.]

miles per hour, and as being Harkness' eighteenth locomotive.[38]

A more complete account was given of the *Pioneer*, the Cincinnati, Hamilton, and Dayton Railroad's new engine, in the *Cincinnati Daily Gazette* of September 25, 1850:

> A. Harkness & Son have just completed a new locomotive engine, designed for the Hamilton Railroad. It is eighteen tons in weight, and embodies many improvements never before introduced in the manufacture of locomotives.
>
> The great objection to outside connecting engines is done away with, being so constructed that the engine runs almost, if not quite as steady, as the inside connecting machines.
>
> The engine has a separate cut off which is beautifully managed and proves itself entirely effectual in its operation. The consumption of fuel is also lessened about one-half . . . .
>
> Much credit is due to Mr. Z. H. Mann, the ingenious draftsman of this establishment, for the arrangement of this beautiful specimen of Western skill. He has been for many years in the East, laboring for the improvement of the locomotive principle and in which he has been entirely successful.

The same paper in its issue of November 5 noted the *Pioneer* had been shipped to Hamilton, Ohio, by the canal and would assist in the construction of the line.

Zadock H. Mann was formerly associated with the Proprietors of the Locks and Canals in Lowell, Massachusetts. In 1838 he was granted U.S. patent 628 (issued March 10, 1838) in association with L. B. Thyng for improvements in locomotive boilers. He apparently left New England in the late 1840's and traveled west, eventually settling in Cincinnati and becoming a designer for Harkness. Mann's earlier connections perhaps account for the marked resemblance of the early Cincinnati engines to those made in New England, notably at Taunton and Lowell. In 1851 Mann became associated with Niles & Co., but was soon succeeded by J. L. Whetstone.

In 1850[39] the Little Miami retired its first locomotive, the *Gov. Morrow*, built by Rogers in 1841, and traded it in to Harkness for a new locomotive of the same name.

---

[38] MARVIN, "Columbus and the Railroads of Central Ohio Before the Civil War."

[39] *Little Miami Railroad: Report for 1850*. The 1857 report states that the *Gov. Morrow* was built in 1852.

23

Figure 9.—PROBABLY THE "CINCINNATI" of the Cincinnati, Hamilton, and Dayton Railroad, built by Harkness in 1851. Scene is Glendale, Ohio, in the 1860's. (*Photo courtesy Mrs. Richard T. Keys.*)

In March 1851 Anthony Harkness, in company with Jacob Strader and T. R. Scowden, Engineer of the City Water Works, left to visit the Great Exhibition at the Crystal Palace in London.[40] During his absence, William took charge and reported to his father in some detail the daily happenings at the foundry. Miraculously, two of these letters have survived. The originals are owned by Mrs. Richard T. Keys of Glendale, Ohio, whose husband was a descendant of Anthony Harkness. These letters, reproduced here in their entirety, give us an intimate picture of the activity at that works in the middle of the last century.

---

[40] *Cincinnati Daily Gazette*, March 21, 1851. The same paper noted his return on June 18, 1851.

Dear Father

April 25 Received an order from Holabird for Two more Boilers. we have three boilers now on hand for him and only one is in a hurry. Finished one boiler to day for shaw & West which we have shipped to them. and drawn on them for the amount as they directed they paying the exchange Paid another enstalment to the Farmers Bank Ky of $500 One Pair of Driving wheels From the Engine Columbus with one of the Tyer broaken which we will weld and return as soon as possible $(4\frac{1}{2}$ ft wheel)

Refused to build a pair of Engines for Wayne for a steam boat. Capt McLean talks something of going on with his boat if he can get good timber. he wants her to come out in October. but I have not made any bargin with him. hav an idea that he will not build. as he has two many (if's in). should he conclude to build would it be better to do it or not. shall wait to hear from you

April 26th Sent an order for 62 sheets of Copper for flues as the other is all worked in flues. The flues are all in the first Hamilton Engine and Driving wheels in the shop ready to be put under.

Have Received 75 tons of Pig Iron from Patterson to day and 12 Tons from Werts & Co. Have finished all the the Hamilton and Mansfield axels for Keck.[41] Commencd the axels for Nashville Road. will put three Lathes on them so as to rush them out. the Pedistals and all of the small Castings are all made

Making Tyer and turning Drivers and Cylenders Plaining side frames for Nasville Engine and Axels for do. Run the [Cincinnati] Water Works Engine for the first time scince puting on new Counter Ballances she a great deal better so much so that we are satisfied not to do anything more in the way of alterations. there is scarsely any jar and what little there is it appears to be equal on boath, that is, when the piston head of the pump strikes the water. and the pump does not fill. Shall let her be now until you return. I think that we cannot better it. The river has been to high ever scince you left for them to put the valve in below it is allready

April 28th Received 44 Tons of Iron from Werts & Co.

Drilling frames for the Tennesee Engine. They are all Plained up and in the shop

Have finished the Bolts and Castings 4 Bridges As for the castings they

---

[41] George Keck. Keck and Davenport (later Keck and Hubbard) were the proprietors of the Fulton Car Works, Cincinnati.

are working on the 6th Bridge   the iron that they last sent from the mill is all very good and will shall not have any more difficulty in working it 29th   The Braces are riveted on the second Hamilton Engine they are fitting Cylinders on do

The Wheels are under the first engine and shall endeauver to get her all done this weeak with the exceptions of Painting they have boath of the houses [cabs] ready to fit on.   It appears to be impossible to push John any on the winding up of an engine   Mr Man[n] and myself have endeauvered to get the first one hamilton engine done by the first of May but it is impossible.   we have lost some 4 or 5 days on this one, shall try and gain it on the next

29   Drawn flues to day 1 sett and Plaining side peices for Truck enough for 4 Engines.   fixing up the Boaring machine to be able to boar the next cylinders commenced moulding cylinders for the Miami engines   Turning Driving wheels for the second Hamilton have 6 Driving wheels cast for the Miami road.   ship to Keck 2 sett of Axels for Tennesee cars

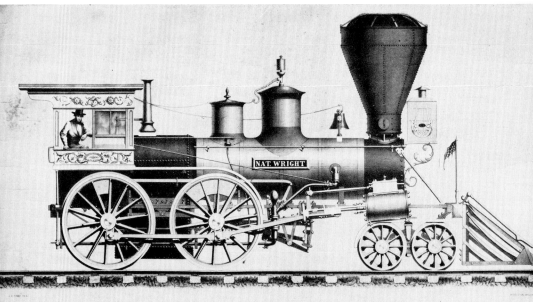

Figure 10.—The "Nathaniel Wright," built for the Little Miami Railroad in 1854 by Moore & Richardson.

Running one of the furnaces on Turnings and the other on Tyers. shall have to run boath on Axels to morrow for Keck as he is in some hurry for the mansfield Cars

Mr. Ham [42] says that the Elk [43] beats them all. Mr Clements [44] seems to be in some hurrey for one of their next Engines

I remain yours
William Harkness

Cincinnati May 2d, 1851

Dear Father

We had steam on the hamilton Boiler to day to see if the Flues and Boiler are tight. they are all right. There is more work to do than they can possible get through with this weeak on her. Tudor is seting up another Boiler for the Hamilton Road. we shipped another Boiler for Mr Holabird to day and another in the yard that is nearly finished for him. We had a severe frost last night that has kiled all the fruit in this secton the Themomiter fell to freezing point. reports from the south say that the Cotton and fruit is kiled by the heavey frost

Sam is geting along with cards. two of them are nearly finished, but a very little more to do to them all They have received the new machienery at the Factory [45]

May 3d Met with a little misfortune to day one of the Tennesee driving wheels broake after it was all turned they were takening out of the lathe it had not the Tier on. Bending the Tire for 3 hamilton and Tennesee Engines. have had some difficulty with the hands about wages let them go and hired others in their places the ballance appear to be better satisfied. as they see that there is no difficulty in geting men to fill their places. I think that there will not be any more difficulty from them

One of the Directors of the Tennesee road is here he wishes to purchase an old Engine if it is possible and also to hurrey us up on the Engine. he would give one Thousand dollars extra if we can get the new engine done in two weeaks which is impossible he has gone to see Mr Clements relative to an old engine I heard from Crawford to day. he wishes to know what the cost of an engine the same as the Fire fly I will tell

---

42 C.T. Ham, master mechanic, Little Miami Railroad.

43 Built for the Little Miami Railroad, April 1851.

44 W.H. Clements, superintendent, Little Miami Railroad.

45 Possibly English machinery, known to have been used by Harkness.

him 8,000.00 dollars he does not say that he wishs to get one only [w]ants to know the price have recast the wheel for the Tennesee engine
May 5 Wiley left this day for Buffalo to attend to Messick[?] & Co suit will begine four days

Mr Man [46] seting the valves for the Hamilton engine she has one coat of paint she will be all finished tomorrow but the painting. the house is fitted on.

Mr More has all of the large bobins turned and the ends grooved the small ones about one third finisched.

We are driving away as fast as possible on all locomotive work Mr Clements has sold the Engine Genl Harrison [47] to the Tennesee Road for 7200. Dollars and the miami Co pays the alterations Ham does all of the work except making the New axels which I have to make. he could just as well as not get $8,000 for her

the Duchess [48] has broaken another Shaft I shall make it for them as there is not much to do in the foundery at the presant time as they are far enough a head in all of the Castings but the Car Wheels which we are making 6 per day. I received a letter from you to day from England.

Mrs Scowden feels very bad to think that she did not receive a letter

The letter you write to me from N York told me to direct my letters to Brown Brothers & Co

The one you wrote to Mothers says Bearing Brothers & Co And the mistake was not discovered until yesterday I shall write to boath houses with this so you will be able to know why I have written to them to correct mis carreying we shall continue to direct to Bearings Brothers & Co. Scince writing the above I have seen Mr Fosdick he says that there is no such firm in England as Brown Brothers & Co. I sent two letters every weeak so you may know

May 6th. The Hamilton Engine is finished but cannot get her out this weeak on account of the boat not ready they let one of the Cars run into her and broake the side in they are repairing. Plaining the pedistals for the third hamilton Engine. Mr Clements says they want 4 Engines 2 — $4\frac{1}{2}$ ft and 2 — 5 ft Drivers as soon as possibley can be made. We are driving away on the work as hard as we can

Accompaning this I send an news paper which gives an account of the

---

[46] Probably Z. H. Mann, draftsman and designer for Harkness.

[47] The *General Harrison*, a 15-ton 4–4–0 built for the Little Miami Railroad in June 1848 by Harkness.

[48] The *Duchess*, a steamboat.

Figure 11.—WILLIAM HARKNESS, 1821–1853, son of Anthony Harkness. (*Photo courtesy Mrs. Richard T. Keys.*)

fate of the S B Webster and the Frost in the South. if you will call at the post office you may find the former letters

I remain yours
William Harkness

In his letter of May 2, 1851, William commented on his method of settling labor disputes. As effective as his dismissal of the workers may have appeared, it did not put an end to such problems. The *Cincinnati Daily Gazette* of March 2, 1852, carried an item entitled "Strikers Struck." The strikers, assistants to the blacksmith who actually wielded the hammer, struck for an increase in wages from $5.50 to $6.00 per week. "This [demand] was not acceded to, and in a short time the strikers—numbering some fifty or sixty collected and proceeded in a body, first to Harkness' foundry and then to Niles', threatening vengeance should any men be employed in their places, and creating considerable disturbance." The police, however, arrived early and induced the men to disperse.

Anthony retired from active management of the foundry in 1851, entrusting this responsibility to William and to ". . . a faithful man who had been in his employ for many years and for whom Mr. H. had formed a strong attachment," one Robert Moore.[49] William, however, as is revealed by his letters, was slow to act without his father's counsel, so Anthony must have continued to be a strong force in the affairs of management.

Although the exact nature of the arrangement is not known, early in 1852 a more formal agreement was reached whereby Moore was made a partner in the new firm of Harkness, Moore & Company.[50]

Robert Moore was born in New York City on December 9, 1805, and migrated with his family in June 1817 to Cincinnati. His father advised him to become a carpenter because: "These were in constant demand anywhere on this broad continent . . . [and] were among the most useful and deserving citizens . . . ."[51] He was apprenticed to Joseph Jones in 1825 and learned the building trade. Moore described his subsequent career:

> In the Summer of 1828 I was employed by the late Anthony Harkness, to assist in erecting his first building, on the north side of Front Street, just east of Lawrence Street, for the manufacture of steamboat machinery and sugar-mills. I then engaged permanently with the said Harkness, to assist in making patterns and drawings for such machinery. This suited me better than my former employment, and I made great progress therein, as draughtsman, foreman, and as assistant superintendent, so that, finally, after a continuous service in this line of twenty-five years, I assumed a controlling interest in the works, and, in conjunction with John G. Richardson (who for a time held an interest in the concern), a lease for twenty years was obtained of the grounds and the buildings, Mr. Harkness retiring with a fortune. Other and outside productive interests had he; but in so far as the estate was accumulated by or through said works, the writer claims to have been largely instrumental, having devoted the best twenty-five years of his young life entirely to the interests of the said Harkness.[52]

---

[49] *Cincinnati Daily Gazette*, May 12, 1858.

[50] This has been determined by checking rosters of the Little Miami Railroad and the Cincinnati, Hamilton, and Dayton Railroad, which as late as December 1851 refer to Harkness & Son but by March 1852 to Harkness, Moore & Co.

[51] MOORE, *Autobiographical Outlines*, p. 34.

[52] Ibid., pp. 35–37.

John G. Richardson had been a foreman at the Harkness foundry from about the time the first locomotives were built there. Edwin Price mentions completing his apprenticeship under Richardson after Mr. Van Loon's death.[53] Practically no further information on Richardson can be found, other than that he was a resident of Newport. The U.S. Census records for 1850 state that he was 35 years old and born in Ohio. The 1860 records give his age as 45 and his place of birth as Bermuda.

The Cincinnati Locomotive Works was formed in 1853 by Moore and Richardson. The new firm took over the Harkness plant and was commonly known as Moore & Richardson.[54] William Harkness apparently dropped out of the partnership at about this time. Several months later the *Cincinnati Enquirer* of November 23, 1853, reported: "Lamentable Suicide—Some surprise was excited in the city yesterday by the announcement that Wm. Harkness, son of Anthony Harkness, Esq., had committed suicide on Monday night [Nov. 21]. He was found dead in his room at his father's Glendale home; the death attributed to morphine."

Locomotives for southern roads were regularly shipped on the river, much as eastern engines were moved via the Great Lakes, since there was no complete rail connection. It was a common sight in the 1850's to see handsome little engines fresh from the erecting shop, their polished brass, vermilion wheels, and bright Russia-iron jackets gleaming in the sun as they moved over temporary tracks to the public landing and an awaiting steamer. The *Cincinnati Gazette* of June 12, 1852, vividly described the scene:

> Look out for the locomotive when the bell rings; ought to be placarded on the public landing, for we certainly saw a locomotive steaming at a good rate from Broadway to Main yesterday morning. A temporary wooden track had been laid down and a new sixteen ton locomotive,

---

[53] Edwin Price reminiscences.

[54] Some confusion exists as to the precise date of the lease of the Harkness foundry by Moore & Richardson. Moore states in his autobiography that he leased the plant in 1852 (pp. 40, 49). The annual reports of the Cincinnati, Hamilton, and Dayton Railroad refer to Harkness, Moore & Company as late as June 1853, and the Little Miami Railroad reports first refer to Moore & Richardson in May 1853. It is therefore assumed that William Harkness retired from the firm in the spring of 1853 and that Moore brought Richardson in as a partner.

Figure 12.—The "Dr. Goodale," Built in 1853 for the Little Miami Railroad. Scene is Xenia, Ohio, in 1865. (Smithsonian photo 42137.)

32

built by Harkness, Moore & Co., for the Memphis Railway [Memphis and Charleston Railroad], was thus working its own passage to the landing of the steamer Memphis.

(Some years later a permanent track was laid along Front Street to the Little Miami Railroad making delivery of engines to connecting lines more convenient.[55])

On June 22 the *Gazette* noted that locomotive number 2 had arrived at Memphis aboard the steamer of that name. The first engine for that road, also from Harkness, Moore & Co., was lost earlier when the steamer *Chickasaw* sank.

In April 1853 Moore & Richardson reported 280 employees with a weekly payroll of $2,200.[56] In May 1854 the firm stated that as many as 300 were employed when the shop was in full production and that it had built a total of 100 locomotives since 1846.[57] In October 1855 the shop's capacity of from 18 to 20 machines per year was announced.[58] Early the next year the *Railroad Advocate* said that over 120 locomotives had been built.[59] A capacity of 50 per year with 400 on the payroll was claimed in 1859.[60] It was also claimed in 1859 that during one previous year the firm had built 23 locomotives and done about $300,000 worth of other work. Only a year later employment was said to be 250. Moore recounted in his autobiography that the payroll occasionally reached $5,000 per week. These figures cannot be verified, since no cashbooks or records of the company exist. Generally, however, most locomotive shops were not worked beyond 50 percent capacity; it can be assumed with some assurance that the management exaggerated the extent and potential of the works for the benefit of the press. The figures given here are not without value, however, for much of this data was screened by such informed and competent persons as Zerah Colburn and does give us an idea of the amount of business transacted. One may also compare the production

[55] *American Railway Times* (April 28, 1866), vol. 18, p. 134.

[56] *Railroad Record* (April 14, 1853), vol. 1, p. 103.

[57] Ibid. (May 4, 1854), vol. 2, p. 148.

[58] *Railroad Advocate* (October 20, 1855), vol. 2, p. 2.

[59] Ibid. (January 12, 1856), vol. 2, p. 1.

[60] *Cincinnati Daily Gazette*, February 5, 1859.

Figure 13.—The "Reuben R. Springer," Built in 1854 for the Little Miami Railroad. (*Photo courtesy E. P. Alexander.*)

of other builders for the period 1850 to 1855. Norris was the largest works of the period, boasting a production of 150 locomotives per year. Rogers averaged about 70 per year and Baldwin about 50. The Portland and the Taunton works, both well-known concerns, averaged 12 and 28 respectively for the years noted. For an obscure midwestern firm little known outside Ohio, the Cincinnati Locomotive Works compares rather well, as it averaged about 16 engines a year between 1850 and 1855.

So far this discussion has been concerned with the vagaries of partnerships and other details of business history, all of which remain somewhat obscure as a result of the ambiguous and fragmentary evidence presently available. The following remarks concerning the mechanical features of the locomotives built by the Cincinnati Locomotive Works (hereinafter referred to as Moore & Richardson) are thankfully less ambiguous since several good accounts of the machines have been uncovered.

The majority of the engines built in this country during the 19th century were the 8-wheel engines known as American types because they were the workhorses of U.S. railroads and were truly the standard and most commonplace type of motive power to be found.

34

Ten-wheel engines were also constructed; they were, in general, intended for freight service and were the second most frequently produced type of engine in the period 1850–1865. Price mentioned in his recollections that a 10-wheel engine built in 1851 for the Nashville and Chattanooga Railroad, the *W. S. Watterson*, had a very heavy outside frame.

The Moore & Richardson locomotives were often said to be copies of Rogers engines, but a close examination of existing photographs has convinced me that this resemblance goes no further than the general arrangement, viz, outside connection and the use of the wagon-top boiler. Other particulars of the Moore & Richardson engines, such as the large single steam dome placed forward on the boiler and the position of the truck, were at variance with the standard Rogers product of the period, being more closely related to the products of Taunton.[61] The association with the "Rogers pattern" is probably due to the fact that the Paterson builders, especially Rogers, were early advocates of outside-connected engines and that those who accepted this arrangement, as opposed to the more conservative New England builders who favored the inside connection, were thus said to copy Rogers. In a manner of speaking, outside connections and Rogers engines became synonymous. Naturally, Rogers built inside-connected engines as well to suit the wishes of the buyer.

The most striking feature of the Moore & Richardson engines was the placement of the truck in front of, rather than centered on, the cylinders. This arrangement probably originated with Griggs in about 1845. It threw more weight on the driving wheels, thus improving traction, but it also extended the rigid wheelbase by moving the center pin of the truck forward another 15 to 18 inches. In addition, the cylinders had to be raised for clearance of the rear truck wheel and then inclined so as to center on the driving axles. Although this arrangement had decidedly gone out of favor by the late 1850's, the 1855 lithograph (see endpapers) shows a spread truck which is placed so that the cylinder must be inclined to clear the rear truck wheel. As is witnessed by the Indianapolis,

[61] A comparison of Taunton's *New England* and *Chamberlin* of the Delaware and Hudson Railroad with Moore & Richardson's *Nat. Wright* reveals the remarkable similarity of the products of these two builders.

Figure 14.—THE LITTLE MIAMI RAILROAD's *William Dennison*, built by Moore & Richardson in 1854. Photograph taken July 4, 1876. (*Photo courtesy Little Miami Railroad.*)

and St. Louis Railroad's number 29, Moore & Richardson maintained this practice until at least 1864 (see fig. 17). There was nothing singular in this arrangement; other builders, too, were prone to favor it. The remarkable fact is that Moore & Richardson was so consistent and devoted to its use.

Moore & Richardson had won a favorable reputation for well-made boilers. As can be observed (see figs. 10, 13, and endpapers), the firm favored a low-crowned wagon top. A large steam dome placed toward the smokebox end of the boiler was preferred in the belief this was the best place to take the steam since the water was less turbulent there than over the firebox. There was, ideally, less chance of priming. Again, many New England builders, as well as Winans in Baltimore, championed this arrangement. A separate safety-valve standard was located at the rear of the wagon top near the cab. Although many builders were abandoning copper boiler tubes for brass or, even better, iron, Moore & Richardson continued to use copper. Copper was originally favored because it was an excellent conductor of heat and, being soft, was easy to flange and expand for a tight fit in the tube sheet. Although it was commonplace as late as 1860, copper gave way to iron when it was learned that iron tubes were not only stronger and cheaper but could evaporate water with nearly the same efficiency. And as coal burning grew more popular, it became necessary to use iron to resist the abrasive action of the coal ash passing through the tubes. The good sense of making the entire boiler assembly of one material, with a view to uniform expansion, is obvious.

As many readers may know, an intense controversy over the relative merits of the suspended or stationary, and the shifting-link motion arose during the 1850's. The stationary-link valve motion was used on many Moore & Richardson engines after about 1853, although by 1860 the triumph of the Stephenson shifting-link motion was undisputed. The following excellent discussion of the use of the shifting link by the Cincinnati Locomotive Works is reproduced from the *Railroad Advocate* of June 16, 1855.

In looking over the report of the Cincinnati, Hamilton and Dayton Railroad, we noticed in the table of engines and engine repairs, that three quite new, Cincinnati built engines, had been fitted with new link motions during the past year. We noticed the fact, as implying a singular and unusual decay of the link-motion, and we suspected that the links had not been case-hardened or that the studs and joints were too small or too short.

We have since learned that these engines were originally fitted with the stationary link, or link suspended from a fixed point; the valve rod being raised and lowered, instead of the eccentric rods. It was found by comparison with other engines having the shifting link, such as that of Rogers' engines, that the latter was more durable, and that it worked steam more economically, so as to use much less wood.

The master of machinery of the road, Daniel McLaren, Esq., therefore altered the link motions on the three engines, all of which were built by the former firm of Harkness, Moore & Co., now Moore & Richardson, and all of which were in all other respects, as excellent machines as any upon the road. (We will say that the road is well supplied with Rogers' engines, and with machines from Taunton, and from the Boston Locomotive Works.)

Since the alteration of the links, which were originally applied as a test of the motion, Mr. McLaren finds the engines require less repairs, and save much wood. The engines, each of 23 tons weight, are now running on passenger trains, using but one cord of wood for each 45 miles run.

The suspended link, or stationary link is but little used. It is applied by one builder at Paterson, who has adhered to it we believe for above

Figure 15.—The Louisville and Nashville Railroad's *Quigley*, built in 1859 by Moore & Richardson. (*Photo courtesy Louisville and Nashville Railroad.*)

Figure 16.—The "Columbus," a 6-Wheel Switcher rebuilt in 1865 at the Pendleton Shops of the Little Miami Railroad. Probably was a Moore & Richardson 4–6–0 of the 1850's.

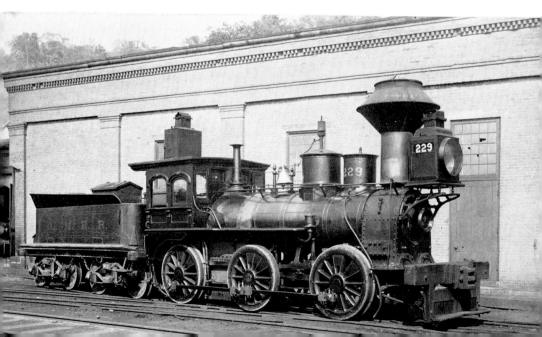

four years. Although it involves more joints than the shifting link, we had supposed that it allowed of a stronger and better suspension. The stationary link maintains a constant lead on the steam port, under every degree of expansion, whereas the shifting link varies the lead, generally from $\frac{1}{16}$th inch at full stroke to $\frac{1}{4}$ or $\frac{5}{16}$ths, in cutting off at three eighths stroke. Yet, although the stationary link keeps a constant lead, always the same, upon the *port*, the mere fact of varying the throw of the valve, varies the period of taking steam, as measured upon the stroke of the piston. With 5 inches throw on the valve and $\frac{1}{4}$ inch lead, the piston will be, perhaps, $\frac{1}{8}$ inch from the end of its stroke when it begins to take steam. With but $2\frac{1}{4}$ inches throw, and the same lead, the piston would be probably one inch from the end of its stroke when it first gets steam to drive it in the opposite direction.

But with the fair trial of the stationary link on the Cincinnati engines, and with the general preference for the other, or shifting, link, as shown in its adoption by every builder who uses the link, with the single exception named, we must believe that the former (stationary) link has some inherent fault. We are well aware that Rogers first attempted the use of the stationary link, and that he soon abandoned it for the shifting link which he has now used for over five years.

We think it would be an interesting subject for investigation and comparison, to have the working of the two plans of links made out and published by their respective builders. If any single builder attaches a loss of 10 per cent. of wood to all his engines, by reason of his unsupported preference for a defective valve-motion, both he and the railroad public should know it.

Despite its failings some roads, including the Little Miami, obtained good results from stationary link motions. William Swanston, formerly of the Little Miami Railroad, commented on them at the American Railway Master Mechanic Association in 1885:

> Some years ago, on the Little Miami Railroad, we had a stationary link on which the lead was constant. Now to get better results [speed] on passenger engines, we increased the lead of the backward motion so as to increase it on the forward motion, and we got good results from that.

Figures 13, 15, and the endpapers will clarify the exact layout of the suspended-link motion. Notice the forward placement of the rocker. The valve stem is extended through the front of the valve box, offering a larger bearing area. Two points of support are provided on the *Quigley* (fig. 15) by a bearing bracket, on which the valve stem extends behind the rocker rather than through the valve box, and by the connection, effected by a Scotch yoke, between the stem and rocker.

In addition to building new engines, Moore & Richardson was active in rebuilding locomotives; it is probable that after 1860 rebuilding constituted a large portion of the firm's activities. The annual reports of the Louisville, New Albany and Chicago Railroad for 1858 and 1859 mention that the locomotives *Planet*, *Comet*, *Meteor*, and *Rocket* were rebuilt by Moore & Richardson. The firm regularly advertised its facilities for rebuilding; an example of such an advertisement can be seen on page 46. (In 1856 the *Railroad Advocate* published two articles describing Moore & Richardson's shops and engines. They are reproduced as Appendix 1.)

Several investigators have been confused by the annual reports of the Little Miami Railroad, which list a number of machines as having been built at its Pendleton Shops, just east of Cincinnati. But for the most part, these were only extensive rebuildings; many Moore & Richardson products were rebuilt there during the 1860's. The *Columbus* (see fig. 16) was reported to have been built in 1865 by Richard Bromley at Pendleton. Yet, it seems unlikely that a progressive master mechanic would build new an inside-connected engine when that type of connection had been out of favor for nearly ten years. The specifications of 4–6–0's built for the Little Miami by Moore & Richardson in the 1850's agree exactly with those of the *Columbus*. A careful study of figure 16 and a comparison of it with other illustrations indicate that the *Columbus* was rebuilt from a Moore & Richardson 4–6–0 built in the early or mid-1850's. Recalling the earlier discussion of the mechanical particulars of Moore & Richardson engines, notice the low-crowned wagon-top boiler, the large steam dome, the safety-valve column, and the cab, particularly in reference to figure 13. The cylinders of the *Columbus* were 15 by 20 inches, the wheels 48 inches in diameter, and it weighed 23 tons.

Anthony Harkness died of cancer on May 10, 1858, after a long and painful illness. A lengthy obituary appeared in the *Cincinnati*

40

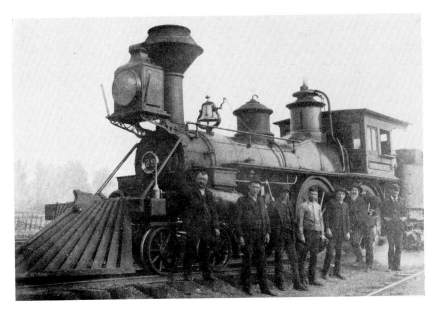

Figure 17.—Engine No. 29, Built in 1864 by Moore & Richardson for the Indianapolis and Cincinnati Railroad, later the Cleveland, Cincinnati, Chicago, and St. Louis Railroad.

Figure 18.—Engine No. 25, Built in 1864 by Moore & Richardson for the Indianapolis and Cincinnati Railroad, later the Cleveland, Cincinnati, Chicago, and St. Louis Railroad. Shown as rebuilt in 1882 with a new boiler.

41

Figure 19.—LETTER SEAL OF MOORE &
RICHARDSON, 1861. (From a letter in
the U.S. National Archives.)

Daily Gazette, from which we have already drawn heavily for information on Harkness' business career. While this account does not
fail to lament with conventional platitudes the passing of an eminent
local business personality, some attempt at objective reporting is
shown in the following appraisal of his character:

> The deceased possessed a mind of great originality and power, but
> it was unpolished and herein consisted his great failing. While he was
> honest in all things, cordial and valuable as an advisor, he was exceed-
> ingly rough in his manner and had but little, if any regard for the feelings
> of others. The latter characteristic was especially prominent, when
> he had occasion to express himself to those he did not fancy.

This is an appraisal that might honestly be applied to many 19th-
century manufacturers, men who were honest, straightforward,
hearty, and in that manner rough and, if so disposed, rude to any-
one who displeased them.

Starting from the humblest beginnings, Harkness amassed half
a million dollars during his lifetime. About 1853 he helped to
found the wealthy residential community of Glendale, located 15
miles northeast of Cincinnati. One can imagine the pride and
satisfaction of Harkness as he commuted on the Cincinnati, Hamilton

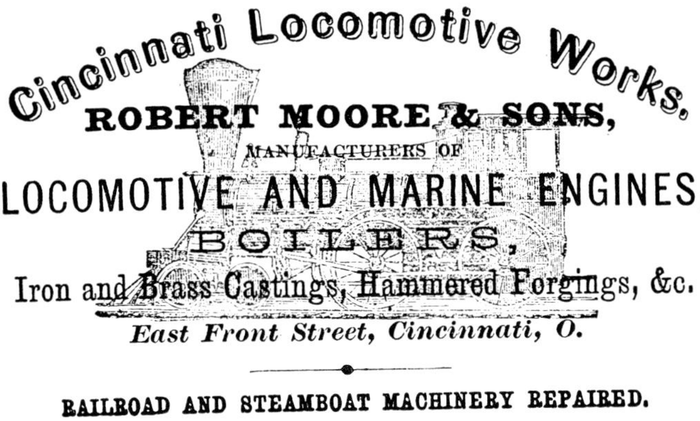

Figure 20.—BUSINESS CARD OF THE CINCINNATI Locomotive Works about 1865. (*Courtesy Mr. Robert L.
Moore.*)

and Dayton Railroad aboard a train more than likely powered by
an engine built in his foundry.[62]

The Civil War brought about a sharp increase in the demand for
locomotives as railroad traffic compounded under the demands of
that struggle. The use of the railroads for military purposes, especially the U.S. Military Railroads, forced the Government to
purchase secondhand equipment and to press nearly every builder
for more engines. Moore claimed that because of this demand his
shop was worked to capacity.[63] It is difficult to determine exactly
for which roads these machines were built, since at about this time

---

[62] An anonymous writer of the early 1860's prepared a little volume called *Trips
in the Life of a Locomotive Engineer* (New York: J. Bradburn, 1863) in which a chapter entitled "Forty-two Miles Per Hour" describes the *D. W. Deshler*, built by
Moore & Richardson in 1854 for the Little Miami Railroad. This account,
reprinted here as Appendix 2, not only presents a lively mechanical picture of
the engine but also conveys the feeling of wonder and pride in the art of mechnaical
technology as it had so rapidly advanced in that day. The trip described is without doubt that from Xenia to Columbus.

[63] MOORE, *Autobiographical Outlines*, pp. 37–38.

Figure 21.—THE "DANIEL McLAREN," BUILT IN OCTOBER 1856 by the
Cincinnati, Hamilton, and Dayton Railroad at Lima, Ohio. This engine
is remarkably similar to those built by Moore & Richardson in the same
period. (Smithsonian photo 45809–A.)

most railroads stopped listing locomotives in their annual reports.
Because the builders' records are not to be found, it is impossible
to calculate just how many were constructed. However, it is
reasonable to assume that none were built for the U.S. Military
Railroads, for no record of Moore & Richardson engines, other
than several purchased secondhand, notably from the Louisville
and Nashville Railroad, can be found on the official Government
roster. Nonetheless, the Cincinnati works enjoyed a prosperity not
seen since the mid-1850's and, indeed, not to be seen again, for at
that time Moore & Richardson was the only midwestern builder in
production, and, with the celebrated eastern shops turning away
orders, its products were much in demand.

Not all war work proved profitable to the firm. In 1863–1864,
Alexander Swift & Company of Cincinnati was awarded a contract
to build two light-draft monitors, the *Klamath* and the *Yuma*, for
river use by the Navy. A subcontract for $261,267 was given to
Moore & Richardson to supply machinery.[64] The two monitors,
along with the other 20 of this class, proved a failure, since, im-
properly designed, they drew too much water to be useful in river
service.[65] Because the design was furnished by the Navy, the con-

---

[64] Preliminary drawings for the engines are in the Civil War Section, U.S. National
Archives, Washington, D.C.

[65] F. M. BERMET, *The Steam Navy of the United States* (Pittsburgh, 1896), p. 483.

44

tractors could not be held responsible. However, a dispute arose, and Moore & Richardson, despite every effort to collect from the main contractor, lost $40,000 on the transaction.

In 1865 Moore & Richardson became Robert Moore & Sons Whether John G. Richardson died or retired is uncertain; however, he is last listed in the city directory for 1864. Efforts to verify the date of his death were not successful, since the oldest death records available for Newport, Kentucky, his place of residence, are for the year 1911. The new firm was composed of Robert Moore and his sons, August, Arthur, and Hamilton.[66]

A contemporary account gives evidence of the difficulties under which the new firm began operations. In a letter (June 29, 1865) to his employer, the M. W. Baldwin Company, William P. Henszey reported on a visit to the Cincinnati Locomotive Works. Henszey, who in 1870 became a partner in the Baldwin Locomotive Works, was visiting other engine builders throughout the country. The letter, preserved by the Historical Society of Pennsylvania, reads in part:

> Was at Moore and Richardsons, saw Mr. Moore also his son who is the general manager. They are entirely closed on locomotive work and have only about 12 men in their establishment. Their shop is about as good an exposition of dirt & confusion as could well be.

Henszey pictured the works in hard times, but rather overstated his case in claiming that locomotive work had been completely discontinued, for the Cincinnati Locomotive Works was advertising for orders in the *American Railway Times* and the *Railroad Record* as late as 1867 and 1868. It may well be that at the time of Henszey's visit locomotive building had been temporarily suspended.

The end of the war found the Cincinnati Locomotive Works in a much weakened condition. Added to the losses on the monitors was another loss of $20,000 resulting from an attempt to establish a sales branch in New Orleans. Too, by the late 1860's the locomotive industry was becoming decidedly specialized as the machine itself became more complex and sophisticated. By then larger freight locomotives such as Moguls and Consolidations were being

---

[66] Information furnished by the late Robert L. Moore, a son of Hamilton Moore, in an interview on December 21, 1959.

built; these were, however, beyond the capacity of the Cincinnati Works, which was well suited only for the building of 4–4–0's and 4–6–0's. The day of the eclectic machine shop which numbered among its products nearly every form of iron goods was rapidly fading.

Although the gunboat and New Orleans ventures weakened Moore's business, the greatest damage was done by the repudiation of credit advances to southern customers before the Civil War. Unable to collect these debts, Robert Moore & Sons was forced into bankruptcy in March 1868.[67] The closing of the Cincinnati Locomotive Works marked the end of the locomotive-building industry in that city.

After the failure of his business, Moore, through efforts of his friends and former business associates, was elected city treasurer for two terms (1869–1874). He lived another 13 years. A year before his death he prepared the invaluable memoir on which I have drawn so liberally for this account. Shortly before his death on May 31, 1887, he presented a copy of this work to the Historical and Philosophical Society of Ohio. It is signed "Very sick, R. M. 1887."

---

[67] MOORE, *Autobiographical Outlines*, p. 41.

Cincinnati Locomotive Works.

ROBERT MOORE & SONS,

CINCINNATI, OHIO,

MANUFACTURERS OF

LOCOMOTIVES,

Marine and Stationary Engines,

IRON AND BRASS CASTINGS,

Boilers, Tanks, &c.

— ALSO —

Repair and Re-build Locomotives.

Mar 3

CINCINNATI LOCOMOTIVE WORK

MOORE & RICHARDSON,

ARE PREPARED TO FURNISH LOCOMOTIVE equal in efficiency and durability to the best Easter manufacture. Also, superior Car Wheels. Shaping an Slotting machines, suitable for Railroad Shops. All kind of Castings. Heavy Forgings by steam-hammer done a short notice. Bridge Bolts cut with dispatch.

Figure 22.—ADVERTISEMENTS OF THE CINCINNATI LOCOMOTIVE WORKS which appeared in railroad journals during the 1850's and 1860's.

Chapter 3

George Escol Sellers'

Grade-Climbing Locomotive

In 1830 Charles B. Vigneols and John Ericsson devised a method to assist locomotive propulsion with a center rail. In this system the engine's regular cylinders and driving wheels are assisted on steep grades by horizontal, smooth adhesion wheels, driven by auxiliary cylinders, which grip a rail fastened in the center of the track. Vigneols, one of the best-known British railway engineers, and Ericsson, a versatile inventor who became associated with nearly every area of applied mechanics, received British patent 5995 for their invention on September 7, 1830. However, the invention, to my knowledge, was never given a practical test. Nearly twenty years later the idea of a center-rail system was revived by George Escol Sellers, who devised probably the most ingenious and advanced design for this type of railway.

This account of Sellers' development of the center-rail system, his unsuccessful attempts to introduce it, and the five locomotives built under his patents is based largely on two sources. The first of these is a remarkable collection of letters dealing with the subject among the Peale-Sellers papers at the American Philosophical Society, Philadelphia, Pennsylvania. This collection contains much of the correspondence between Sellers and the other individuals involved with grade-climbing locomotives. The second consists of the Minute Books of the Panama Railroad, U.S. National Archives, Washington, D.C. The information revealed by these two firsthand sources refutes much that has been written in the past

47

Figure 23.—GEORGE
ESCOL SELLERS,
1808–1899.

about locomotives and, in certain instances, contradicts the recollections of Sellers and his associates.

The Sellers family was well known in Philadelphia for its mechanical skill and ability.[68] George's grandfather, Nathan, and father, Coleman (the elder), produced a great variety of machinery in their shop including fire engines, papermaking and textile machinery, leather hose, and presses for several of the U.S. mints. It is not surprising, then, that George became interested in the study of the mechanical arts.

At the time of his father's death in 1834, George and his elder

---

[68] Information on the Sellers family and the early activities of their machine shop is drawn from an excellent series of nearly 40 articles by George E. Sellers entitled "Early Engineering Reminiscences," which appeared in the *American Machinist* from 1884 to 1895. The obituary of G. E. Sellers that appeared in the same journal on March 20, 1899, was also consulted. See *Early Engineering Reminiscences (1815–1840) of George Escol Sellers* (U.S. National Museum Bulletin 238, edit. Eugene S. Ferguson; Washington: Smithsonian Institution, 1964).

Figure 24.—ILLUSTRATION of the type of locomotives built for the Philadelphia and Columbia Railroad by Sellers in 1836. (From the Philadelphia city directory of 1839.)

brother, Charles, assumed the management of the family business at Cardington, a few miles west of Philadelphia. Late in the summer of that year the Board of Canal Commissioners asked the Sellers company to build locomotives for the Philadelphia and Columbia Railroad and for the Allegheny Portage Railroad. Both lines were part of the huge public works which was to connect Philadelphia to western Pennsylvania by a system of canals and railroads. Locomotives were in great demand at this time; however, few shops in this country were equipped to build them. Since the Sellers brothers were now building heavy machinery for iron furnaces and rolling mills, they had machine tools of a size sufficient to handle locomotive work. Two 4–2–0's, named the *America* and the *Sampson*, were completed in 1835 and 1836.[69] The depression of 1837 closed

---

[69] For detailed descriptions of these machines see the *American Machinist* (October 31, 1885, and November 7, 1885), vol. 7, pp. 4 and 1.

their shop, however, and it is believed that no other locomotives were built by the Sellers in Philadelphia except the *Atlantic*, constructed for the Baltimore and Susquehanna Railroad between 1838 and 1839.

George moved to Cincinnati in the fall of 1841 and opened a lead-pipe works there in association with Charles and the wealthy Cincinnati merchant, Josiah Lawrence.[70] In about 1844 the lead-pipe business was sold, and the firm of C. Sellers & Company opened the Globe Rolling Mills and Wire Works. The *Cincinnati Gazette* of February 22, 1845, reported that the building, located on southwest Front Street, was completed and that the machinery was being installed. Strap rail, iron rods, and telegraph wire were produced along with other mill products. In January 1845, shortly before the rolling mill was completed, George Sellers obtained U.S. patent 3882 for a drop hammer. The hammer was attached to a lifting bar which was gripped by two smooth friction wheels. The friction wheels could tighten or release their hold on the bar, so that the hammer fell when the wheels were released. Leaf springs and toggles pressed the friction wheels together. In October of that year the hammer was tested at the shop of Miles Greenwood, a manufacturer of hardware and stoves.[71] One hammer was also put to work at the Globe Rolling Mill. Sellers stated many years later that his experience with the drop hammer led him to develop his grade-climbing locomotive.[72]

It would be easy to assume that Sellers merely adapted the mechanical arrangement of his hammer for the center-rail system, but several factors contributed to the invention. While building the two locomotives for the Philadelphia and Columbia line, George and his brother patented a scheme for using the weight of the train

---

[70] Anthony Harkness built the machinery for this shop (*American Machinist*, December 19, 1889, vol. 12, p. 2).

[71] *Cincinnati Gazette*, October 9, 1845.

[72] *American Machinist* (August 29, 1895), vol. 18, p. 684.

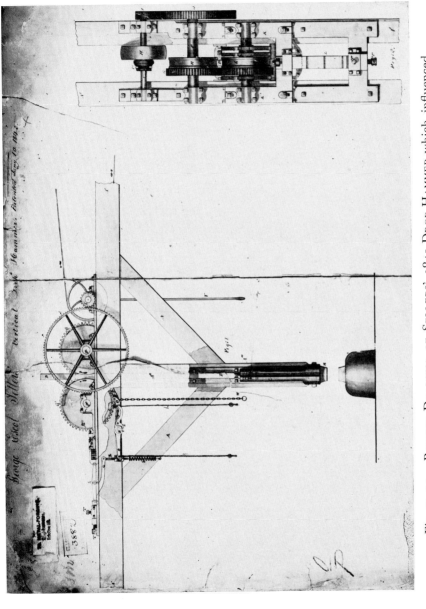

Figure 25.—Patent Drawing of Sellers' 1845 Drop Hammer which influenced him in the development of his grade-climbing locomotive. (From U.S. patent 3882, January 10, 1845.)

51

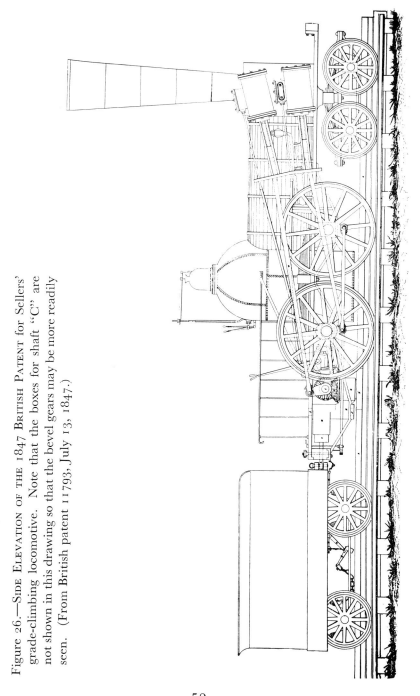

Figure 26.—SIDE ELEVATION OF THE 1847 BRITISH PATENT for Sellers' grade-climbing locomotive. Note that the boxes for shaft "C" are not shown in this drawing so that the bevel gears may be more readily seen. (From British patent 11793, July 13, 1847.)

to increase the weight on the engine's driving wheels.[73] As we shall soon see, the use of the train's weight to increase the locomotive's adhesion was one of the cardinal principles of the grade-climbing engine. At the same time, 1835–1837, George proposed a scheme to increase the locomotive's traction by use of a special rail with rollers running under the driving wheels. The rollers would be under pressure, pulling the driving wheels hard against the top surface of the rail. The fallacy and impracticability of this idea was quickly pointed out, and it was never tried.[74] Each of these ventures shows that Sellers was concerned with locomotive adhesion sometime before he began work on the drop hammer or the center-rail engine. Too, there had been much local interest in grade-climbing locomotives since the opening of the cog railway on the incline of the Madison and Indianapolis Railroad in 1845. At about the same time that Harkness was building his first locomotives (1846–1847), Sellers' idea for a grade-climbing locomotive began to mature.

His engine operated on the level as a conventional machine. When on a grade, two smooth gripper wheels, mounted under the locomotive and driven by an auxiliary set of cylinders, engaged a center rail, thus adding their tractive effort to that produced by the driving wheels. The weight of the train became a positive rather than negative factor, since it pulled upon a drawbar connected to the gripper wheels by means of toggle-levers; thus, the heavier the train and the steeper the grade, the tighter the grip of the adhesion wheels on the center rail. Although Sellers did not admit it, this principle was not a net gain, for the friction on the journals supporting the gripper wheels increased with the load of the train. The toggles and the levers by which the adhesion wheels were

---

[73] A patent was issued to Charles and G. E. Sellers on May 22, 1835 (no number); the restored patent is now in the U.S. National Archives, Business and Economics Records Section. The specifications are in vol. 23, pp. 429–30, of restored patents; the drawing number is X 8839. In this scheme a lever fastened to the rear of the boiler or frame of the locomotive (4–2–0's only), with the driving wheels' axle as the fulcrum, would be attached to the train in such a way that the weight of the train, when ascending a grade, would pull on the lever and cause the engine to tip up, thus throwing the weight normally carried by the leading wheels upon the drivers.

[74] See SELLERS, *Improvements*, p. 17, for a description.

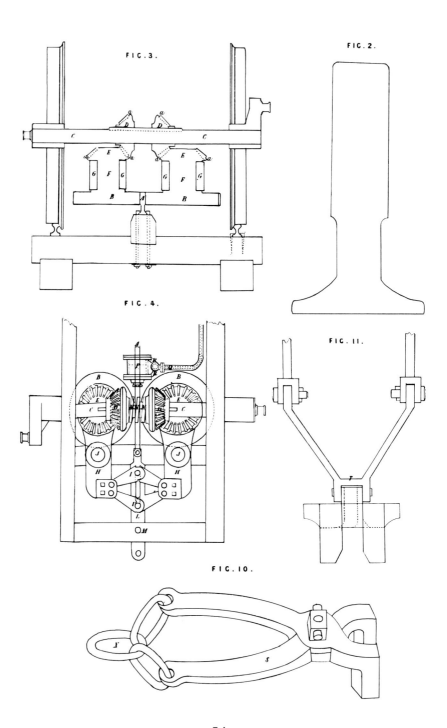

FIG. 3.

FIG. 2.

FIG. 4.

FIG. 11.

FIG. 10.

54

attached to the main frame served another important function: they allowed the locomotive to vibrate laterally between the running rails. If the gripper wheels had been rigidly attached to the running gear it would have put an undue strain on the center rail.

On the level, at stations or fueling places, where the weight of the train would not be available to help start it, a steam cylinder or heavy spring could apply the necessary pressure on the gripper wheels. The steam cylinder could also apply more pressure if the wheels should slip while pulling a light train up a grade. The weight of the train could also be used to advantage when descending a grade, in this case as a brake. The toggles were made double so that they could press the adhesion wheels against the center rail whether the train was ascending or descending the grade. In this way the auxiliary cylinders of the gripper wheels, working forward or in reverse under light steam, could hold the train back on heavy gradients. The center rail, it should be noted, was not to be laid the entire length of the line, but only at heavy grades, stations, and other places where extra traction was needed.

It was claimed that since adhesion was achieved by the pressure of the gripper wheels rather than by the *weight* on the drivers alone, a 10- or 12-ton engine on the center-rail plan would produce the same tractive effort as a 20-ton machine of the ordinary pattern. This was the chief advantage claimed for the scheme. The same results could be achieved by rack or cable, but Sellers was quick to point out the complexity and expense of such systems. A more detailed accounting of the principles and claims of this invention

←

Figure 27.—DETAILS OF THE CENTER RAIL (fig. 2), adhesion machinery (figs. 3 and 4), and safety clamp brake for the center rail (figs. 10 and 11) as shown in the 1847 British patent. In fig. 4 of this plate, "P" is the steam cylinder used to increase pressure should the draw-bar weight prove insufficient; "KKK" is a coil spring to keep the lever engaged; "HH" are the massive levers connected to the frame by pins "J-J" which support the adhesion wheels; "I-I" are the toggles. "A" is the center rail, "B" the adhesion wheels, "L" the draw-bar which connects the train and toggles, "M" a pin by which the draw-bar could be held stationary. The distortion of "B" in figure 4 is due to the drawing lying across the fold of a page.

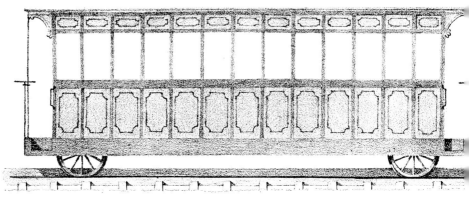

Figure 28.—PIONEER SYSTEM LOCOMOTIVE AND COACH. The adhesion wheels are located near the center of the locomotive at "B." (From Sellers' *Improvements*.)

can be found in the patent specifications. An examination of figures 26 and 27 will also assist in the understanding of this mechanism.

Sellers' ideas went beyond his adhesion locomotive: he devised a new plan and system for the construction of railways. To consolidate the country and improve its internal communications, the United States was faced with the prospect of building a vast system of railways over a rugged and unsettled land where little capital was available. Sellers felt that the trend to heavy locomotives and permanent ways was not the best solution of the problem. Light, cheap railways and motive power sufficient to climb heavy grades would be satisfactory until the country was more fully developed. When the land was settled and a prosperous economy established, more permanent roads could be constructed on the conventional pattern. Little grading needed to be done for Sellers' center-rail locomotive because it could, except in the most extreme cases, traverse the natural rise and fall of land. Lines could be direct; they could climb over rather than bypass or tunnel through natural obstacles. Since his locomotives were light, the roadbed, rails, and bridges could also be light. The Sellers engines could, without fear of damage, be run over the light railways faster than heavier conventional machines. All these factors combined to offer a method

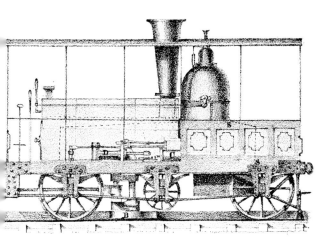

for building cheap railways into the remote and unsettled areas of the country, notably the West, which might otherwise have to wait many years for a convenient connection to other parts of the country. The idea for light railways was carried to its extreme in a scheme Sellers came to call his Pioneer System. (It is discussed at length in his pamphlet *Improvements in Locomotive Engines and Railways*, published in Cincinnati in 1849.[75]) The locomotive intended for the Pioneer System and illustrated in figure 28 depended entirely on the center rail for propulsion. The wheels were merely idlers to carry and guide the engines.

The Pioneer System, inspired by the plank-road craze, was a cheap form of the center-rail plan in which wooden rails were used with the lightest equipment possible. All three rails were to be of wood, which eliminated the necessity of importing English iron. The author of *Observations*[76] claimed the country was being drained of its gold to pay for imported rail. Building all new roads on the Pioneer System would end this injustice, permitting the new gold

---

[75] An anonymous pamphlet on this subject, *Observations on Rail Roads in the Western & Southern States, and of the Introduction of the Pioneer System for their Construction, with Remarks on the Importation of Foreign Iron*, was reprinted in 1850 from the *Cincinnati Chronicle and Atlas*. The author may have been Sellers or one of his associates interested in promoting the invention. Of two copies known by me to exist, one is in the Library of Congress, in a group of pamphlets entitled *General Technology*, vol. 30, and the other is at the Historical and Philosophical Society of Ohio, Cincinnati.

[76] See preceding footnote.

from California to remain here and finance internal improvements. Such roads could be built from local timber in backwoods areas for about one third the regular cost, or at an estimated $677 to $1,223 per mile. Again bridges and grading would be cheaper because of center-rail propulsion.

Sellers recommended that the Pacific Railroad be built on the Pioneer System. Costs and construction time would be reduced by half, he asserted, and California bound to the Union many years before otherwise possible. The practicability of the proposal is questionable; still the country was to wait nearly twenty years for the first transcontinental line to open. Whatever the merits of the Pioneer System, it was not used for the Pacific or any other line. With no one willing to test the System, Sellers did not continue to promote it and thereafter devoted his energy to perfecting the center-rail idea for main line use.

A British patent for Sellers' grade-climbing locomotive was issued in the name of Alfred V. Newton[77] in July 1847. United States patent 5367 was granted on November 13, 1847, and French and Belgian patents were also secured. An association composed of Miles Greenwood, William M. Hubbell, and John P. Foots was then formed in order to promote Sellers' invention. The group held the patent rights with Sellers and financed him for nearly two years, during which time he obtained a second U.S. patent, prepared a pamphlet, and in general devoted all his time and energy to the adoption and establishment of the center-rail locomotive.

In order to demonstrate the utility of the invention, it was decided that a practical test should be made. Full-size machinery would have cost more than the association was willing to spend, so a large model locomotive was built. It was produced in the shops of the Globe Mill by Sellers, his younger brother Coleman, and a young Cincinnati mechanic, John L. Whetstone, in the spring of 1848. It was described in *Improvements* as follows:

> The boiler is made of copper; the cylindric part, or body, is 12 inches in diameter, flues 21 inches long and 62 in number, 55 being 11–16th of an inch and 7 being 1–2 inch in diameter; heating surface in fire-box and

---

[77] Alfred V. Newton was probably associated with *Newton's London Journal of Arts and Sciences*. Numerous patents of foreigners were issued to A. V. and William Newton of this publication.

flues, 23 square feet, and grate surface, 160 square inches. The cylinders are 3 inches in diameter, 4 1-2 inch stroke; driving wheels, 14 inches in diameter, 1 inch truck wheels, 8 inches diameter; wheels on the tender and car, the same as truck wheels. The tender is mounted upon 6 and the car upon 8 wheels. The car will seat 16 grown persons comfortably, but has, at times, carried 32. Width of track, 18 inches; weight of engine, 1300 pounds; weight of tender, with fuel and water, 450 pounds, car 400 lbs.—in all, 2150 pounds.

In the model engine, the adhesion wheels are 6 inches in diameter, and the gearing is arranged, so that when the engines which drive them, make the same number of revolutions per minute, as the engines which operate the side driving wheels do upon the level road, the engine and train shall advance at a speed of one third of that on the level. In this manner the same volume of steam is used upon the gradient, as upon the level, and the power is increased in proportion to the diminished speed of the train. The rapid exhaust of the auxiliary cylinders, maintains the intensity of the blast whilst ascending the grade, and consequently enables the boiler to generate an adequate supply of steam, so that a deficiency of steam cannot occur on the gradients, which, would be the case, if the blast were diminished by the smaller number of revolutions per minute, and consequently less rapid exhaust.

In constructing the model engine, some improvements were introduced, which will be beneficial to the ordinary locomotive. The arrangement of the eccentric and valve gear is such as to permit the upward and downward motion of the axle, on which they are placed, without disturbing the position of the valves, which is found to be so detrimental to the working power of the best constructed locomotives, when running over rough places in the road. This arrangement is not shown in the drawing, but is so simple, that it may be rendered intelligible by a description. The eccentrics are surrounded by bands, which are made to slide up and down with the motion of the eccentric, (or with the play of the axle,) in yokes, which are made to vibrate or slide in such manner, that the sides of the yokes are always parallel with the sides of the pillow-blocks, in which the axle boxes slide, so that the valve motion taken from these yokes is never disturbed, in whatever part of the yoke the eccentric band may be placed. [This valve motion is shown in fig. 29.]

The working of all the parts of the model far exceeded my expectations. I refer particularly to the workmanship displayed in its construction, and will mention one fact to the credit of those employed in building it. The cylinders are of brass, with pistons with metallic packing, and so well fitted, that, on turning the cranks of the upper pair of cylinders, in a direction opposite to the set of the valve gear,

59

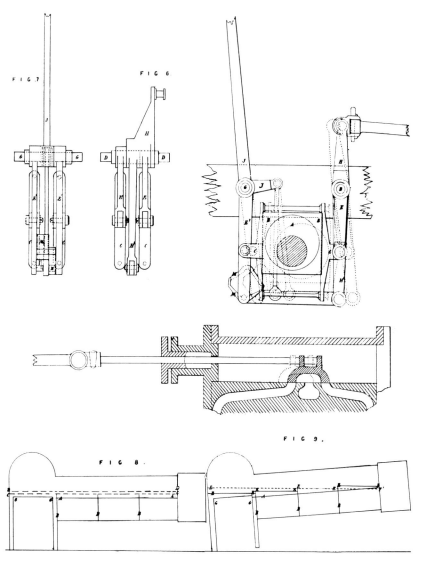

Figure 29.—Details of the Valve Gear (figs. 6 and 7) and the boiler (figs. 8 and 9) as shown in the 1847 British patent. Neither the boiler nor this valve gear with its curious Scotch yoke was shown in the U.S. patent. Note the use of baffle plates in an attempt to maintain a constant water level in the boiler.

air will be compressed into the boiler in 5 revolutions, to impel the engines with the adhesion wheels and bevel gear back through 4 revolutions.[78]

By April 1848 the mechanical work on the model was nearly complete,[79] and on May 20 it was given a public trial in Cincinnati. Plans were made to exhibit the model in this country and in Europe. However, despite repeated references in the letters of the Sellers brothers and in various journals to a projected display of the model in Europe—Coleman and Whetstone studied French with the idea of showing the model in France—this was apparently never done.[80] According to a typewritten note in the Peale-Sellers papers, the trip to France was cancelled because of the revolution of 1848. As late as February 1849 Sellers talked of going, but he became so involved with the Panama Railroad that he abandoned the idea of introducing his invention in Europe.

Late in June the model was taken to New York City by George, Coleman, and John Whetstone. A track was set up on a vacant lot at Fourth Avenue between 22nd and 23rd Streets. Sellers stated many years later that he had been invited to display the model by the president of an association planning to build the Panama Railroad, Thomas W. Ludlow, who offered the lot for a demonstration.[81] Horatio Allen, consulting engineer to the Panama Railroad, suggested that Sellers' plan be considered because of the great cost and difficulties anticipated in building the road across the Isthmus.

Invitations brought many of the most eminent railroad men in the country, including John B. Jervis, Benjamin H. Latrobe, John Brandt, and William Norris, to New York for the trials. The first showing was made in early August 1848. The little locomotive was put through a number of severe tests. Two passenger cars were loaded with 33 persons, a total of about 5,000 pounds, and

---

[78] Pp. 7–9.

[79] Peale-Sellers papers (exact date not given).

[80] Peale-Sellers papers, vol. 3, pp. 14–15; *Scientific American* (August 19, 1848), vol. 3, p. 380; *Farmer and Mechanic* (August 24, 1848), vol. 2, p. 402.

[81] G. E. Sellers to C. K. Lord (vice president, Baltimore and Ohio Railroad), February 16, 1893, Peale-Sellers papers. The letter contains Sellers' reflections on his grade-climbing locomotive.

pulled up a grade of about 3 degrees (276 feet per mile). The locomotive had some difficulty in starting with this heavy load when relying entirely on the regular driving wheels, but when the center rail was put into use, it ascended the grade without difficulty. The most dramatic test came about by chance: the regular drivers were accidentally put in reverse, but the power of the gripper wheels was so great that it propelled the locomotive and car up the grade while the driving wheels slipped, turning vainly in the opposite direction.[82]

Whetstone described other incidents that occurred during the New York showing in a letter to George Sellers:

> In the exhibition yesterday, not being able to find my man Friday nor anyone else in his place, I had my hands full. I had besides the engineering department, to do the talking, and to act as fireman and policeman. It being Saturday afternoon, and the crowd of boys was very great, and they were unusually troublesome, so much so that they had come very near doing very serious mischief. While the engine was passing off the lower turn-table to go to the depot some boy shifted the turn-table so as to throw the truck off and the engine went into the house minus a truck, it was very lucky that the hind end of it was heavy enough to keep it from tilting down, otherwise it might have been ruined entirely. I soon corrected matters, however, and everything went on as well as ever.

In another paragraph of the same letter he notes:

> Yesterday about noon Mr. Norris sent a note to Mr. Woodward [Jabez M. Woodward] requesting an exhibition of the engine to some gentlemen from New Orleans and elsewhere, appointing 5:00 o'clock as the time. Mr. Randon[?] came there with Mr. Woodward and Mr. Jas Lee brought a friend with him, Mr. Norris came afterwards bringing Judge or General McNeal [Major William G. McNeill] one of the U.S. Engineers who at present resides in this city. Mr. Norris says you must see Mr. McN on your return. I suppose probably he thinks he will take an interest. Mr. Norris certainly takes a great interest in this matter. He has already brought 6 or 8 gentlemen to see it and talks to them very much in the same strain that he did to you. What his reasons or motives may be I do not know.[83]

---

[82] SELLERS, *Improvements*, p. 10.
[83] September 10, 1848, Peale-Sellers papers.

The motives of William Norris in encouraging Sellers are open to question. A compulsive promoter, in need of a new project to promote, Norris had returned to this country in 1848, having been forced to abandon his connection with the Vienna Austrian locomotive works because of the revolution in that country. Finding himself not welcome at the Norris Works at Philadelphia, then operated by his brother Richard, he attached himself to the most spectacular railway project of the period, the Panama Railroad. Through Norris, Sellers was drawn into the enterprise. (The ruthless schemes of several of the men attempting to organize this company, their maneuvers to swindle one another, and their questionable dealings with the government of New Granada [now Colombia] are complex and obscure [84] and of no interest here other than as relating to Sellers.)

Early in 1848, shortly before becoming U.S. consul in Venezuela, Colonel John P. Adams was sent to New Granada by C. H. Todd and his associates to secure a grant from that government for the Panama Railroad.[85] Adams was offered a one-third interest in the company but, apparently feeling that he could do better by acting as his own agent, kept his activities a secret from Todd and his company. Adams seems to have received some assurances of obtaining the grant; at least he represented matters as such to Norris. Norris was especially interested in getting the rights to this charter because of the prospect of heavy traffic across the Isthmus as a result of the discovery of gold in California. He persuaded Sellers to join him in forming a company to obtain the charter from Adams and to build the first ocean-to-ocean road in the New World. Sellers agreed, since he felt that he could surely secure a trial of his adhesion locomotive if he were an official of the Panama Railroad. No better advertisement for his invention could be obtained than a demonstration on this railroad, whose progress and operation would command international attention. On October 25, 1848, Norris

---

[84] See JOHN H. KEMBLE, *The Panama Route—1848–1869* (Berkeley: University of California Press, 1943), for a discussion of the intrigue among Todd, Aspinwall, and Sellers' associate, Jabez M. Woodward.

[85] C.H. Todd to W. Norris, December 30, 1848, Peale-Sellers papers. Todd had earlier (November 1847) lost the U.S. Mail contract for the Panama route to Aspinwall.

and Sellers offered Adams $250,000 to transfer the New Granada grant to a company they hoped to form.[86] The precise events immediately following this offer are not clear, but by means of political influence William H. Aspinwall succeeded, independently of Adams, in obtaining the prized charter in December of that year.[87] Norris summarized his unhappy situation upon their failure to secure the Panama grant in the following letter:

New York Jany 22^d 1849

My Dear Sellers

I must beg you to excuse this short letter, written just on the eve of sailing for Chagres, I only decided two days ago to accept an office on the Expedition to make the Survey & Location – and I go out as one of the Chiefs of Division—I will write you a full history of the whole matter in detail, For the present, I can only state in general, that General Herrau [?] did not keep his promise to Adams, but gave the Grant to Aspinwall—Aspinwall then made his own selection of Engineer, and offered me a secondary place—which I could not accept, until I was satisfied that my chance for obtaining the Contract, would be benefitted by my going in person—For this purpose I go & for no other—Manrow [?] was left out of the organization——

You may depend on receiving a long letter in detail & believe me to be

Yours very sincerely
William Norris [88]

During the exhibition of the model locomotive in New York Sellers obtained testimonials [89] from many of the more prominent engineers who viewed the model. Horatio Allen commented enthusiastically on the value of the center-rail system. (Considering the unusual articulated double-enders he built for the South Carolina Railroad in the early 1830's, Allen was not likely to be disturbed by innovations in locomotives.) His endorsement, along with the general respect and prestige he enjoyed in railroad circles, had

---

[86] Copy of agreement among Sellers, Adams, and Norris, October 25, 1848, Peale-Sellers papers.

[87] G.E. Sellers to C.H. Todd, February 14, 1849, Peale-Sellers papers.

[88] W. Norris to G.E. Sellers, January 22, 1849, ibid.

[89] These were later published in *Improvements*, pp. 20–26.

much to do with the eventual adoption of the plan by the Panama company. He was to maintain this view throughout the period of Sellers' association with the Panama Railroad, and he remained loyal to both Sellers and his system until the final abandonment of the center-rail scheme.

Not all the testimonials were unqualified; while most were favorable, certain objections were voiced. Benjamin H. Latrobe, chief engineer of the Baltimore and Ohio Railroad, believed that the locomotive would be useful in special cases where very steep gradients were encountered. He did not advocate the general adoption of the center-rail system, believing its real merit lay only in its remarkable ability to climb grades which could not be handled by ordinary engines. He suggested that engines based on the Sellers plan be built solely as grade-climbing machines, without regular cylinders or connections and fitted only with driving machinery for center-rail operation. Thus they would be confined to sections of the road with steep grades and used as pusher engines. Possibly Latrobe was considering the heavy Appalachian grade the Baltimore and Ohio was yet to overcome. He also questioned the possibility of a level crossing of two railroads with the center rail. Sellers admitted the impracticality of such crossings, but suggested that because of their inherent danger level crossings should never be constructed in the first place. The general opinion seemed to be that Sellers had developed a useful invention which might be used to great profit on roads with steep gradients to overcome. In August the Committee on Arts and Sciences of the American Institute of New York inspected the model and awarded Sellers a gold medal for his improvement.[90]

The model remained on exhibit in New York at least through September 1848. During this time Sellers was conferring with railroad officials in New York, Washington, Baltimore, and Philadelphia, hoping to find someone sympathetic to a full-size trial of his invention. Throughout this period Sellers kept in close contact with Horatio Allen and the Panama Railroad. A detailed survey had not yet been made,[91] and Allen continued to express fear that the

---

[90] *Farmer and Mechanic* (August 24, 1848), vol. 2, p. 402.

[91] A preliminary survey made early in 1848 by J. L. Baldwin and John L. Stephens indicated the lowest pass through the Isthmus to be 300 feet above sea level.

Figure 30.

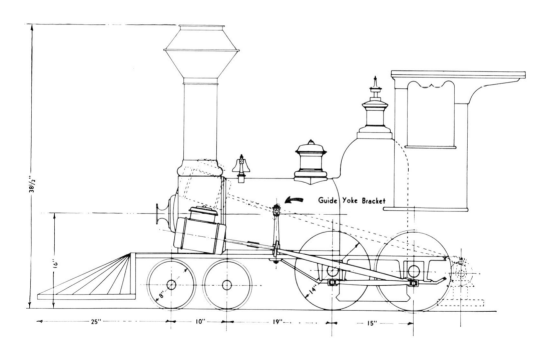

NOTE: Auxilary Cylinders & Grade Climbing Machinery
Shown in Phantom.

66

Figure 30.—THE 18-INCH GAUGE MODEL exhibited in New York
City in 1848 by G. E. Sellers, now on display at the Ohio
State Museum in Columbus. The grade-climbing machinery and auxiliary cylinders have been removed. (*Photo
courtesy Ohio State Historical Society.*)

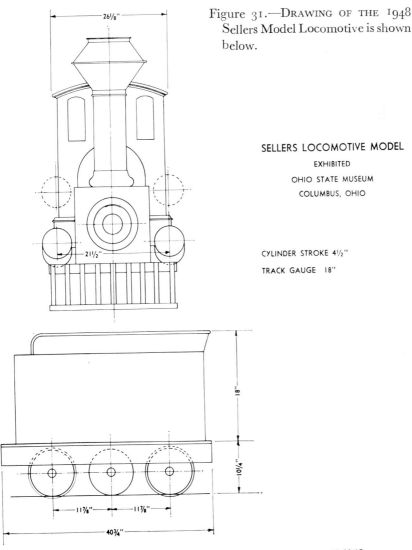

Figure 31.—DRAWING OF THE 1948
Sellers Model Locomotive is shown
below.

SELLERS LOCOMOTIVE MODEL

EXHIBITED

OHIO STATE MUSEUM

COLUMBUS, OHIO

CYLINDER STROKE 4½"

TRACK GAUGE 18"

NO SCALE

DRAWN: APRIL, 1961   J. H. WHITE

67

line would have to pass over extraordinary grades through the almost impenetrable jungle of Panama. The idea soon became established in the minds of the Panama officials that the Sellers plan might provide the only feasible means of opening their line.

Early in September 1848 George wrote to his brother Charles in Cincinnati:

> My prospects of succeeding with my locomotive is good and improving—although it is of necessity slow work or rather an uphill business, I have not yet made any attempt to form a company, but have devoted all my expectations[?] to having it adopted for some road.[92]

He added that he expected to stay in New York another five or six weeks, but he did not return to Cincinnati until late December.

The model was returned to Cincinnati and stored on the third floor of Greenwood's foundry.[93]   Sellers later described as much as he knew of the fate of model:

> I left it in the attic of the Greenwood Works and I supposed it was destroyed at the time those works were burned [1852] but shortly before Greenwood's death [1885] he informed me that it went to Columbus, Ohio and was used to run a work shop. He gave names, but I was unable to trace it. At that time Greenwood's mind had so failed that his account was not coherent. I have since heard, that the engine with the grip wheels taken off was exhibited at the Cincinnati exposition as the work of some of Greenwood's apprentices. It may still be in existence.[94]

It is, although its true identity has been completely obscured since it was given to the Ohio State Museum, Columbus, in 1903–04 by L. B. Davies. While visiting the museum in August 1959, I noticed this curious model on display and was immediately struck by the antiquity of its mechanical arrangement: the Bury boiler, inclined cylinders, round side rods, and the attachment of the main rod to the rear driving wheel. These features, so reminiscent of the early Eastwick and Harrison 4–4–0, indicated that the model was patterned after a machine built before 1850. Little was known of the model, but it was believed to be connected with a pilot invented by

---

[92] September 2, 1848, Peale-Sellers papers.
[93] J. L. Whetstone to G. E. Sellers, August 5, 1859, ibid.
[94] See footnote 81 above.

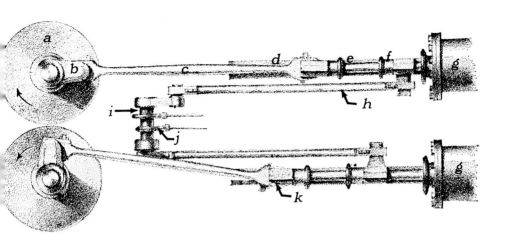

Figure 32.—THE ARRANGEMENT OF THE INSIDE-CONNECTED adhesion machinery is shown here in plan view: (a) adhesion wheel, (b) crank, (c)
connecting rod, (d) crosshead guide, (e) piston rod, (f) secondary crosshead, (g) cylinder, (h) secondary connecting rod, (i) counter shaft, (j)
eccentric, and (k) main crosshead. (From Sellers' *Improvements.*)

the donor. Little further thought was given to the matter until I
came across Sellers' statement that the model had been shipped to
Columbus. This and the striking similarity of the model described
by Sellers to the model on exhibit at the Ohio State Museum immediately suggested that they were the same.

Investigation revealed that the basic dimensions of the model are
identical to those Sellers lists in his *Improvements* (pp. 7–9): 8-inch
truck wheels, 14-inch driving wheels, 18-inch track gauge, 4½-inch
stroke, copper boiler, brass cylinders, and a tender mounted on 6
wheels. An examination of the machine in 1961 further revealed
that the valve gear, fitted to the rear driving axle, is identical to that
described in Sellers' British patent of 1847 (see fig. 29). Also, the
column attached to the frame to which the crosshead guide yoke is
fastened has an identical block and threaded hole, now plugged, at
its upper extremity. This can quickly be seen by referring to figures
30 and 31. This hole was the point of fastening for the crosshead
guide yoke for the upper cylinders. The smokebox, a square brass

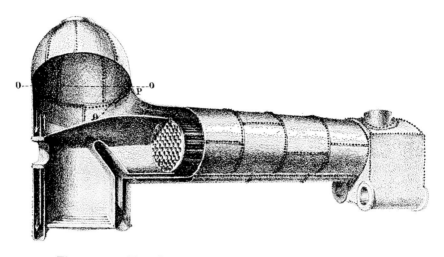

Figure 33.—The Improved Boiler Was Designed to keep the water level at line o–o. Note the water leg projecting over the grates back toward the fire door and the water course running full length on the underside of the boiler. (From Sellers' *Improvements.*)

casting, has plugged holes indicating the position of the upper cylinders. Certain vestiges of the grade-climbing machinery are evidenced by several studs on the frame behind the rear driving wheels and by the continuous drawbar on the tender with a slot cut in the end sill.

The true identity of the model had been lost through a misunderstanding at the time it was donated, and for nearly 60 years its construction has been credited to the donor, L. B. Davies. The correction of this error would probably be of little comfort to Sellers, considering the bitter disappointments the grade-climbing locomotive and every venture connected with it held for him.

When Sellers returned to Cincinnati in December 1848, he received the news of the failure of his scheme to obtain the Panama charter. Nevertheless, he still hoped to secure a test on that road through the assistance of Allen, or Norris, if the latter should win the contract. In any event others had voiced serious interest in its adoption during the New York trials. To further these hopes, he set to work to improve the invention itself and to devise new means of broadcasting its virtues.

70

Figure 34.—THE ORIGINAL PATENT MODEL for Sellers' 1850
patent, now in the collections of the U.S. National Museum.
The main and side rods which connected the outside cylinders to the driving wheels were not incorporated in the
model because they were not considered necessary to
illustrate any part of the invention. (Smithsonian photo
46884–A.)

It was at this point that Sellers began to write *Improvements*, the
pamphlet referred to previously. He planned to distribute it among
interested railway men, not only to promote the invention but also
to illustrate certain improvements he had made in the design since
the first patent. The pamphlet has 26 pages of text and 4 plates,
and was undoubtedly one of the most elaborate and well-written
prospectuses in its field prepared during that period.

With his new design Sellers wished to silence the most common
criticisms of the center-rail engine, namely the use of gears and the
inadvisable attachment of the auxiliary cylinders to the upper part
of the smokebox. The weakness of this arrangement is obvious:
the cylinders would eventually come loose from the thin sheet-iron

71

sidewalls of the smokebox. Furthermore, the reciprocating action of the four cylinders, placed at such a distance from the locomotive's center line, would cause it to lurch from side to side. To correct these difficulties, the cylinders were fitted inside the frame under the smokebox. The gripper wheels, toggles, etc., were brought forward to a position directly before the axle of the forward driving wheel. In short, the basic arrangement of the adhesion wheels, toggles, etc., as originally conceived in the 1847 patent was retained, but the drive was changed from an outside-geared to a direct inside connection. The engines of each gripper wheel were connected by means of a countershaft, fitted with cranks set at right angles, and other necessary rods so that neither engine would stop on dead center. The eccentrics for driving the valve gear were also mounted on this secondary shaft. A clearer picture of this mechanism can be gained from figure 32.

Other advantages can be credited to the new inside-connected design. The extremely long main rod previously required to connect the bevel gearing and the auxiliary engine was eliminated. More important, the rigid wheelbase was shortened by placing the adhesion wheels in front of the front driver. In the old plan, when the adhesion wheels were engaged with the center rail, the rigid wheelbase extended from the center pin of the leading truck to the center of the adhesion wheel's shaft. This increased the wheelbase an estimated 30 inches. In the new plan the rigid wheelbase would naturally be shorter, being no more than that of an ordinary 4-4-0 (from the truck's center pin to the rear driving wheel), and would thus allow the engine to traverse sharper curves.

Sellers also introduced certain reforms in the design of the boiler which would keep water on the crown sheet at all times, whether the boiler was level or inclined and thus insure more complete fuel combustion and better circulation of the water. In the model engine the boiler had to be kept almost full of water when ascending or descending steep grades so as not to burn out the crown sheet. This caused priming and resulted in loss of power and strain on the machinery. During the first test of the model in Cincinnati with the grade set at 5 degrees, this actually occurred. Sellers was careful not to attempt a grade of more than 3 degrees during the New York trials. Even with these difficulties, however, there was undoubtedly a sufficient supply of steam. The strong draft created by the blast of the four cylinders would keep a lively fire burning. The improved boiler design is described in greater detail in *Improvements:*

Figure 6, Plate II, [fig. 33] represents a form of boiler which admits of any desired length, whilst it gives, at the same time, an increased fire surface. The general internal construction of the boiler is shown by a vertical section through the fire-box. The fire-box is separated from an expanding chamber by a water-bridge, which compels the flame to pass through a contracted space before entering the expanding chamber, where the gaseous products of combustion are ignited previously to entering the flue-tubes. By this device, coal or coke can be used without injury to the flues. The water line is marked at 0,0. The flues occupy the entire circle of the body of the boiler, thus giving an increase of heating surface. The top of the body of the boiler is made slanting at an angle somewhat greater than that of the greatest elevation to be overcome. This is done for the purpose of giving free passage into the steam chamber to the steam evolved at the forward end of the boiler, when ascending a gradient. If the body of the boiler were made cylindrical, the steam would rise to the forward end, forcing the water from it into the dome, until the water in the cylindrical part had come to a level with the point P, at the dome end; thus exposing the forward ends of some of the upper rows of flues to injury from excessive heat. By confining the water line to the diameter of the steam chamber, or dome, the variations of the level of its surface, from the elevations or declivities of the road, are very slight.[95]

The Sellers boiler, with the exception of the water leg and the circulating courses, is modeled very closely on the Bury boiler. Sellers had, of course, become familiar with this type of boiler and its hemispherical dome when building locomotives for the Philadelphia and Columbia Railroad in 1836. At that time it was the most popular design among U.S. builders, but by the early 1850's it had gone out of favor, ironically at the very time Sellers obtained his boiler patent. It is interesting to note, however, that the Lima Locomotive Works revived this general scheme in the 1880's by applying it to Shay-geared locomotives. Called a "boot" or "T" boiler, it was used because Shay engines operated on industrial lines with steep grades, and the large dome, as Sellers

---

[95] Pp. 11–12.

Figure 35.—Top View of the 1850 Patent Model with boilers removed, showing the mounting levers, the counter-shaft and its cranks set at 90°, the four crossheads, and the connecting rods. (Smithsonian photo 46884–C.)

had observed, was able to keep water over the crown sheet regardless of the grade.

Rearrangement of the machinery from an outside to an inside connection had been proposed in the columns of the *Scientific American* as early as the time of the New York trial.[96] Sellers apparently had worked out this modification by the time the new boiler was devised, for he included both ideas in his petition (February 20, 1849) for a new patent. It was issued on July 9, 1850, as U.S. patent 7498. The original patent model is now in the collections of the U.S. National Museum and is pictured in figures 34, 35, and 36.

---

[96] (August 19, 1848), vol. 3, p. 380.

Figure 36.—Bottom View of the 1850 Patent Model showing the toggles, adhesion wheels, and the draw-bar. (Smithsonian photo 46884–B.)

While Sellers was working on his pamphlet and patent in Cincinnati, the affairs of the Panama Railroad were developing rapidly. On April 7, 1849, the company was incorporated in the State of New York. In October of that year a preliminary contract was tendered to George M. Totten and John C. Trautwine for construction of the road.[97] About five years earlier, both had successfully surveyed a canal through tropical Colombia and were therefore not without qualifications to undertake the present project. Norris, who had undoubtedly suffered especially from the heat and pests during the Panama survey because of his age and obesity, was no longer an active contestant for the contract. He

---

[97] Panama Railroad, executive committee minutes, October 12, 1849.

dropped out of the railroad to open a gold mine in the area which later proved a failure. Sellers was not, however, without influential friends. Trautwine had been a personal friend since early school days, and Horatio Allen, who remained as consulting engineer with the Panama company, continued to favor the center-rail system.

Writing to John L. Stephens, president of the Panama Railroad in July 1850, Trautwine urged the company to use "Sellers' plan of a locomotive to overcome high grades instead of making heavy excavations." [98] Some years later, however, in the *Journal of the Franklin Institute*, Trautwine disclaimed any advocacy of the Sellers plan beyond ". . . the use of such engines on that road, further than as a temporary resort, while making the summit excavations; and even that idea originated with the Board of Directors, who were determined to leave nothing unprovided for that could expedite the work. I expressed to them the conviction that no such precaution was necessary . . . ." [99]

Sellers consulted with the directors of the Panama Railroad during the summer of 1850. In early August they offered to hire him as a mechanical engineer and began negotiations for the use of his patents. [100] On September 6, 1850, Sellers accepted an annual salary of $3,500 as engineer and $10,000 in stock ". . . for use of his patent right for locomotives in case the same should be used and prove satisfactory . . . ." [101] Sellers was to supervise and inspect the construction of all locomotives, cars, pile drivers, and other equipment. He was also to inspect the timbers and pilings which were to be precut and treated here before shipment to Panama. (About five miles of line was to be built on piles and cribbing.) His main interest, of course, was in seeing that his locomotive be given a trial, and he quickly became irritated by his many administrative duties, which included among other things buying a used locomotive from the Reading as a construction engine.

George set his brother Coleman and J. L. Whetstone to work in Cincinnati on the general arrangement drawings for the Panama

---

[98] Ibid., July 23, 1850.
[99] (January 1868), vol. 55, p. 15.
[100] Panama Railroad, executive committee minutes, August 9, 1850.
[101] Ibid., September 6, 1850.

locomotives. By late September, Coleman reported that the drawings were nearly complete and that after he had rendered the elevations in color they would be forwarded to New York.[102] These drawings were submitted to several builders for quotations. Only three submitted bids, and the Norris Brothers of Philadelphia testily stated, "Mr. Sellers plan of Engine is so different from the usual kind, that no fixed price can be named . . . ."[103] Apparently no builders were interested. Nothing could have delighted Sellers more; for several weeks before soliciting the bids he had commented to his brother that he would like to quit his job with the Panama and act as an independent contractor for the locomotive.[104] Coleman answered, "I hope you can make some arrangement to get the building of the locomotives in your own hands . . . ."[105]

On November 1, 1850, Sellers drafted a contract between himself and the railroad. He proposed to build three locomotives, not exceeding 12 tons, for $9,750 each. The first was to be delivered in New Orleans by June 1, 1851, the other two by July 1 of that year. The company was to advance him, in periodic payments, one half the total price, the balance to be paid on completion.[106] On November 19 the following entry was made in the Panama Railroad executive committee minutes:

> Mr. Allen also read a proposition from Mr. Sellers offering to contract to build three locomotives in Cincinnati deliverable at New Orleans, and stated to the Company that from the great demand for work and the large orders now on hand by the different locomotive builders together with the fact of Mr. Sellers being largely interested in the success of his plan, his opinion was that the company would get a much better style of workmanship by contracting with him.[107]

The company agreed to Sellers' terms, and a formal contract for three locomotives and tenders at the proposed price and delivery dates was signed a few days later.

---

[102] Coleman to G.E. Sellers, September 25, 1850, Peale-Sellers papers.

[103] Norris Brothers to W.H. Townsend, November 4, 1850, ibid.

[104] G.E. to Coleman Sellers, September 27, 1850, ibid.

[105] Coleman to G.E. Sellers, October 10, 1850, ibid.

[106] Paper marked "1st rough draft of Locomotive contract Nov. 21, 1850," ibid.

[107] Executive committee minutes, November 19, 1850.

Certainly Sellers desired a test of his locomotive, yet his willingness to build these machines at such a low price is not altogether understandable, since a conventional 8-wheel engine cost about $8,000 or $9,000. Several motives are possible: other builders may have quoted a high price to build the complex engines because they were not in want of other profitable work; Sellers may have feared that the high price of engines built by another firm would cause the Panama to drop his plan or that the engines might be poorly built and fail mechanically. Therefore, he was willing to build the machines at little or no profit to insure a successful demonstration of the invention. He may also have speculated that any losses would be made up by income from the $10,000 in stock. (As it turned out, the Panama, once it opened in 1855, paid 12 to 20 percent dividends for many years.)

Since Sellers had withdrawn earlier from the Globe Rolling Mill to devote full time to the grade-climbing engine, he had no place to build the locomotives in Cincinnati. Not troubled by this, he began to purchase machine tools in the East. He directed his brother to secure a small stationary engine and to prepare their small shop and residence, located on Sixth Street between Cutter and Linn Streets in Cincinnati's West End. The small work could be done there; the heavy pieces he would farm out to other shops in the city. He wrote to Coleman with enthusiasm regarding these preparations—"I have all the arrangements matured in my own mind" [108]—and requested that the shop and drafting room be in good order so that upon his return work could begin at once on the working drawings and patterns.

Before leaving for Cincinnati he ordered the flues and the driving wheels. From Lepech & W. Manns of Philadelphia he ordered 420 iron flues 8 feet long and of $1\frac{3}{4}$-inch outside diameter; they were to be shipped as soon as possible via New Orleans. He decided to use iron rather than copper tubes, claiming that they would save about $300 per boiler and had been found to be better than copper or brass tubes by tests held on the Erie and Reading Railroads.[109]

---

[108] November 22, 1850, Peale-Sellers papers.
[109] Ibid.

Six pairs of 42-inch wrought-iron driving wheels complete with axles and crankpins (three pairs to be fitted with eccentrics) were ordered from Finch & Welley of Liverpool, England, through Gilmore, Blake, & Ward of Boston. Sellers instructed the agents that prompt delivery was important and that the drawings and gauges he would furnish must be followed exactly, since other components of the locomotives would be built while the wheels were being made in England and all must fit together in the final assembly. The first set of drivers, two pairs of wheels and axles, were to be sent as soon as finished, for the first locomotive was scheduled for an earlier delivery than the others.[110]

After Sellers' return to Cincinnati early in December, work began in earnest on the drawings, contracts were let for building the boilers, and arrangements were made with Greenwood, Globe, and the Niles brothers to make the larger castings and forgings. Now, after nearly three years of tedious negotiations and frustration, work had actually begun on a full-size locomotive. Sellers felt his invention would soon receive the worldwide recognition he had planned for it.

Before discussing the construction of these engines further, however, several misconceptions regarding them should be pointed out. The Globe Works and, more often, Niles and Company have been credited with building these machines, but, as noted, each produced only parts, not the entire engine. It has also been claimed that four or five locomotives were built; we know that only three were constructed. One of the most inaccurate comments was made in the usually sound *Railroad Advocate* (August 15, 1857, p. 42), where it was reported that the Sellers grade-climbing locomotives had five cylinders each. The most common error is the use of the side elevation of the 1847 patent drawing to represent the Panama engines. It is a fair representation of the model, except that the drivers are too large in diameter, but it is in no way an accurate illustration of the Panama locomotives, which were fitted with only one pair of outside cylinders and no gears. The Panama engines are more accurately depicted by the 1850 patent drawing

---

[110] G.E. Sellers to Gilmore, Blake & Ward, November 25, 1850, ibid.

Figure 37.—THE CENTER-RAIL locomotive as pictured in Sellers' *Improvements*. It is the most accurate representation of the engines built for the Panama Railroad in 1851–1852. Meant to illustrate the improved inside-connected center-rail engine, the plate was made nearly two years before the Panama engines were built. (Smithsonian photo 46943–C.)

or, even better, figure 37. The working drawings of these engines were retained by Sellers until 1893, when they were lent to the Baltimore and Ohio Railroad for its exhibition at the Columbian Exposition. In recent years I have made several attempts to trace these documents, but without success.

By the spring of 1851 substantial progress had been made on the first locomotive. The erecting was done in several rude sheds hastily built on a vacant lot near Sellers' little machine shop. In late March or early April the first set of driving wheels arrived from England.[111] The first engine was finished sometime in July, for on August 3, 1851, Coleman wrote to his sister Anna: "Tomorrow we try the locomotive for the last time before shipping it . . . when we tried it last it worked beautifully better than engines usually do at their first trial."[112]

While the first engine neared completion, abandonment of the

---

[111] John Finch to G.E. Sellers, March 25, 1851, ibid.
[112] Peale-Sellers papers.

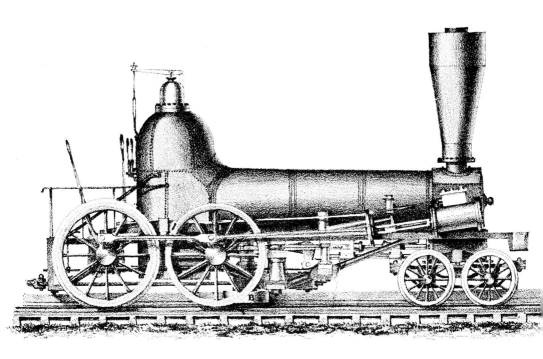

center-rail plan was being considered by the executive committee of the Panama Railroad.[113] Sometime after Trautwine's resignation from the project in November 1850[114] a new survey revealed an easier grade, one that eliminated the necessity of using the Sellers locomotives. Sellers was unaware of these developments and proceeded to New York in early August to be on hand before the locomotive arrived for the inspection. Coleman informed his brother that it had been shipped on August 9 and should arrive in Cleveland two days later.[115]

Upon the locomotive's arrival in New York, the officers of the Panama Company insisted on a trial of the machine and refused payment. Allen reported that the engine was sound and that a trial could be held only at great expense and inconvenience, since

---

[113] Panama Railroad, executive committee minutes, July 25, 1851.

[114] Trautwine wrote the editors of the *American Railroad Journal* (January 11, 1851, vol. 7, pp. 27–29) that he did not resign because of ill health, as often reported, but because he could not agree with the directors.

[115] Coleman to G.E. Sellers, August 10, 1851, Peale-Sellers papers.

the locomotive had been partially disassembled and packed for shipment to Panama. He confided to Sellers his suspicions of the new management and his belief that since the company had changed its plans the demand for a trial was only a pretext to avoid payment.[116] One misfortune seemed to follow another. Whetstone, who had proved so indispensible during the design and construction of the engines, became ill and had to quit temporarily just as Coleman was trying to complete the two other locomotives. Coleman reported that he was doing his best to carry the work along, but that the mill and lathe had broken down and he had been obliged to borrow money to meet the shop's payroll.[117] A settlement was finally reached whereby Sellers was paid $7,111.[118] The locomotive, its adhesion machinery removed, was shipped to Panama as an ordinary engine. Sellers must have returned to Cincinnati with little heart to complete the two other engines for not only would he suffer a serious financial loss on the contract, but he would not see his invention demonstrated.[119]

In late December 1851 Sellers was told to complete the other locomotives, removing the adhesion machinery, and ship them to Panama as soon as possible. They were needed to assist in the construction work. Shipment to New Orleans was delayed by ice in the river, but on February 13, 1852, the *Cincinnati Gazette* reported: "A splended locomotive and tender, built by Sellers & Co., of this city, was shipped yesterday on the *Cincinnatus* for New Orleans. It is intended for the Panama Railway, and is called the 'Isthmus'." It was carried to Panama aboard the schooner *Fannie*, leaving New Orleans on March 14, 1852. A few days later the third engine departed on the brig *Mechanic*.[120] For what comfort it may have

---

[116] G.E. to Coleman Sellers, August 28 and September 8, 1851, ibid.

[117] Coleman to G.E. Sellers, August 26 and 31, 1851, ibid.

[118] G.E. to Coleman Sellers, September 10, 1851, ibid.

[119] In his letter to C.K. Lord (see footnote 81 above) Sellers recalled that John L. Stephens, president of the Panama company, had promised him a public demonstration in this country even though the engines would not be used in Panama, and that this was not done because of Stephens' death. However, Stephens died on October 10, 1852, sometime after the initial rejection of Sellers' plan.

[120] Panama Railroad, executive committee minutes, March 30, 1852.

Figure 38.—JOHN C. TRAUTWINE, 1810–1883, did much to promote the grade-climbing locomotive and recommended its use on the Panama and the Coal Run Improvement Railroads. (From the *Railroad Gazette*, Oct. 5, 1883.)

afforded the builder, Francis Spies, secretary of the Panama company, also reported that " . . . from the information I have been able to gather about the engines built by Mr. Sellers, they perform well but are too light to draw much of a load being only about 12 tons." [121]

Sellers took the Panama company to court, at first winning a judgment on the $10,000 in stock promised him for use of the patent. This decision was reversed, and he was without funds to continue to fight in the courts. As late as February 1853 he petitioned the directors for payment; however, the company replied, and not without justification, that nothing was due him since the center-rail system had not been used.[122] Financially exhausted by this venture, Sellers was forced to close his shop. Coleman and Whetstone went to work for the Niles brothers, who had by this time added locomotive building to their business. Whetstone accepted tools from George in return for back wages.

In order to repay his debt to Niles for the work they had done on the Panama engines, Sellers undertook much of their drafting work at his home, regularly working until 10 o'clock in the evening. To support his family he became, in addition to performing some engineering duties, a traveling salesman for Niles: he received $100 for each locomotive sold. Indeed, his prospects for the moment were dim.

At the time the center-rail locomotive seemed entirely dead, J. C. Trautwine became associated with the Coal Run Improvement Railroad, to be built in the coal mining area near Mount Carmel, Pennsylvania. Unusual grades, as great as 150 feet per mile, were encountered where the road was to cross "the summit which divided the waters of the Susquehanna from those of the Delaware near the Catawissa stage road."[123] At first an inclined plane or a switchback was thought necessary to overcome this heavy grade. Trautwine proposed the center-rail system, a suggestion readily received by the company's directors. In a letter to Sellers of November 9, 1853, [124]

---

[121] Ibid.

[122] Ibid., directors' minutes, February 15, 1853.

[123] Abstract of *First Annual Report of the Coal Run Improvement Railroad*, in *Railroad Record* (March 9, 1854), vol. 2, pp. 22–23.

[124] Peale-Sellers papers.

Trautwine enthusiastically reported the adoption of the grade-climbing locomotive.

Boyers Tavern
Unexplored Regions
Wed. Nov. 9, 1853

Dear Sellers,

I avail myself of a shockingly rainy day, to drop you a line. I am busy with a corps of assistants, exploring a new line for our Road, through the mountains of our Coal region. The Little Schuylkill Co. propose to make a road to meet us, in case the new route is adopted. In order to finish my first explorations before the heavy snows begin, we breakfast and set off before sunrise, and don't knock off until sunset. In some 2 or 3 weeks, I hope to see enough to enable me to form a tolerable approximate estimate of the costs of the 2 routes. This will be much longer than the other; and the heaviest grade will be at one end, about 120 feet to a mile ascending landing; the next will descend towards Market. So in either event your engines will come into play; indeed I told the Directors that I should advise their use, if we had ascending grades of over 30 feet per mile. I think it probable that much of the winter will be lost in pow-wow-ing between the two Companies, and that they will not get to work in good earnest before spring. *They are all in favor of the center rail.* I have been applied to by several Engineers for copies of your book, and have of course distributed willingly. One of the copies has gone out to the Rocky Mtns. to the U S Engineers in charge of one of the Pacific RR Routes; and the matter is daily attracting more & more attention. In order to stir our people up, I have told them that you would require about a years notice for getting up the first pair of Engines. They have just issued a Call for the first instalment of stock, and I hope will move soon in earnest. They are determined to press forward as soon as they can decide upon the new route. The old route would have been let out, and work commenced already, had not the Little Schuylkill Co. suggested making a road in connection with the route I am now at. So that although we *seem* to be standing still just now, such is not the fact.

Give my best respects to your family, and to Mr. Whetstone and believe me.

Truly yours,
John C. Trautwine

In the first annual report of the Coal Run Improvement Railroad, Trautwine stated in his comments as chief engineer that he was gratified by the progressive spirit of the board of directors and their confidence in his judgment on the capabilities of the Sellers grade-climbing locomotive. "I cheerfully encounter the doubts of the unbelieving in view of the prospective credit that must attach to the success of the measure." [125]

A contract was made with Niles & Company to build two locomotives. They were to be 0–8–0's weighing 30 tons each, more than twice the size of the little engines built for the Panama. The following table lists the pertinent dimensions [126] (except in the two respects noted, the machines were identical):

Pairs of drivers	4
Diameter of drivers	44 inches
Diameter of cylinders	16 inches
Length of stroke	20 inches
Weight of engine with fuel and water	60,592 pounds
Weight on drivers with fuel and water	60,592 pounds
Weight of tender with fuel and water	29,680 pounds
Grate area	3,164 square inches
Length of combustion chamber	44 inches (60 inches on the second engine)
Number of flues	144
Length of flues	138 inches (122 inches on the second engine)
Outside diameter of flues	2 inches

Little else of a mechanical nature is known of the engines, for no drawings or other illustrations have survived. The original drawings were retained by Sellers as late as 1893, but their ultimate fate, as with the drawings of the Panama engines, is unknown. As recalled by Alexander Mitchell, the well-known locomotive designer of the Lehigh Valley Railroad, the adhesion machinery

---

[125] *Railroad Record* (March 9, 1854), vol. 2, pp. 22–23.

[126] Data supplied by W. E. Lehr, superintendent of motive power, Lehigh Valley Railroad, from its motive power record books. The dimensions listed are those of the machines when in service on the Beaver Meadow Railroad.

was inside-connected and the drivers were outside-connected.[127]

It is possible that more than one set of adhesion wheels was used, for Sellers said in *Improvements* (p. 18): "Where locomotives of great power are to be constructed . . . it will be necessary to employ two or more sets of adhesion wheels, either geared together, or connected with each other by the usual series of connecting rods, commonly used in locomotives."

Work began in February or March 1855, for the March 29 issue of the *Railroad Record* stated that Niles was "now building" two Sellers patent locomotives for a road in Pennsylvania.[128] The engines were built under the direct supervision of Sellers and Whetstone. As work began on the engines, Trautwine reported to the directors of the Coal Run company:

> The two Sellers locomotives, contracted for with Messrs. Niles & Co, of Cincinnati, will, as I am advised, be ready for delivery by the stipu-lated time, viz: In April next.
>
> I have appraised Mr. Sellers that our road will not be ready to re-ceive them at that date; but I anticipate no difficulty in making arrange-ments with him for storing the machines for us.[129]

One of the engines was given a public trial in Cincinnati in late August 1855, by which time both locomotives had been completed.[130]

While the locomotives were under construction, the Coal Run company was experiencing considerable difficulty, which led to its failure before the railroad itself was complete. Trautwine at the same time lost his right arm in an accident connected with the road's construction. The locomotives were sent East at just about the time of the failure. J. Snowden Bell recalled having seen as a

---

[127] SINCLAIR, *Development of the Locomotive Engine*, p. 317. An identical description was reported to C. E. Fisher by Henry F. Calvin (see *Bulletin* no. 42, Railway and Locomotive Historical Society, p. 28). Calvin mistakenly reported that the engines were originally equipped with cog gearing rather than the smooth center-rail machinery.

[128] Vol. 3, p. 86.

[129] *Second Annual Report of the Coal Run Improvement Railroad*, 1855, p. 10. Traut-wine's report was dated February 6, 1855. This annual report can be found in the extensive collection of such material in the library of the Association of American Railroads, Washington, D.C.

[130] *Railroad Record* (August 30, 1855), vol. 3, p. 422.

boy two locomotives of this description ("large for that time") in the West Philadelphia yards of the Pennsylvania Railroad about 1856 or 1857. One was called the *John C. Trautwine* (later renamed the *Champion*) and the other, he thought, the *Defiance*.[131] The machines were sold at a sheriff's sale and, according to J. C. Trautwine:

> . . . at very low prices to other coal companies. As none of the officers of these companies understood the principle or mode of action of the engines, they did not even make a trial of their capabilities, although it might have been done, and the grand problem satisfactorily demonstrated to all, for a few hundred dollars. As the worthy president of one of the companies himself complacently informed me, We are all practical men on this road, and don't believe in thy gimcracks. The center-rail machinery was accordingly taken off and melted up for castings . . . .[132]

Either the coal companies referred to in the foregoing quotations were subsidiaries of the Beaver Meadow Railroad or they sold these locomotives to the Beaver Meadow before 1864. The Lehigh Valley Railroad, when it absorbed the Beaver Meadow in 1864, acquired the *Defiance* and the *Champion*. They were numbered 19 and 20 and reportedly did good service as switchers until about 1878.[133] Sellers, in a letter to C. K. Lord, recalls hearing reports that they were regarded as ready steamers.

Sellers' interest in the center-rail locomotive came to an end with the collapse of the Coal Run Railroad scheme and, with it, any likelihood of a demonstration. After spending nearly eight years in its promotion, building five full-size locomotives, and enduring the numerous travails previously mentioned, it is small wonder that in his memoirs he dismissed the entire subject of the grade-climbing

---

[131] J. C. Trautwine, Jr., typewritten note, January 28, 1924, Peale-Sellers papers. Trautwine recalled these facts from a conversation with and an earlier letter from Bell, who had related that the machines were built for the Panama Railroad, which we know to be untrue.

[132] J. C. Trautwine to Henry Morton, December 6, 1867, in *Journal of the Franklin Institute* (January 1868), vol. 55, p. 15.

[133] W. E. Lehr to the author, November 4, 1960; data supplied by Clinton T. Andrews.

locomotive by saying: "I do not propose to say anything of the labor, care, and anxiety that invention caused me. It is too un-ending a net to get into."[134]

EPILOGUE: THE ULTIMATE SUCCESS OF THE CENTER-RAIL SYSTEM

If the center-rail system had begun and ended in Sellers' ordeal, there would be little justification, beyond its antiquarian interest, for the foregoing lengthy discussion. A brief outline of the later development and adoption of this idea, however, will illustrate its utility when applied to special purposes such as mountain railways.

John B. Fell took out a number of patents on a center-rail loco-motive in England between 1863 and 1869. His locomotive used four adhesion wheels, which were pressed against the center rail by springs. Sellers' arrangement of toggle levers was a far more refined plan for pressing the adhesion wheels against the center rail, and the Fell arrangement was repeatedly criticized in the technical press, especially in *Engineering*,[135] as anything but an improvement on the Sellers plan. Nevertheless, a line on the Fell plan was opened in 1868 through the rugged Mont Cenis pass between France and Italy. It was built as a temporary measure while the Mont Cenis tunnel was under construction. The road functioned well enough mechanically, but closed in 1871 when the tunnel was opened. Several other Fell lines were opened in other parts of the world, notably New Zealand; they did good service for many years. One line, the Snaefell Railway, on the Isle of Man, was still in operation in 1964.

While the Fell line over Mont Cenis was being built, Zerah Colburn, Coleman Sellers, and J. C. Trautwine were quick to point out George Sellers' previous invention.[136] Still, the ultimate success of the third-rail system apparently proved of little satisfaction to Sellers, who did not choose to comment further or in any way challenge his imitator.

---

[134] *American Machinist* (August 29, 1895), vol. 18, p. 684.

[135] Numerous references to the Fell system are found in this journal from 1867 to 1869.

[136] Their comments appeared in the *Journal of the Franklin Institute* and *Engineering* in 1867.

The French are reported to have considered a center-rail system for battlefield railways during World War I. As late as 1920, it was claimed that such roads could be quickly and cheaply built and would do good service in hauling supplies and heavy artillery over the rough terrain of the front lines.[137] The similarity of these claims to those expressed for the center-rail system by Sellers should be readily apparent to the reader.

---

[137] "Traction on Heavy Gradients by Means of Auxiliary Adhesion," *Bulletin of the International Railway Association*, Brussels (January-February-March 1920), vol. 11, pp. 272–284.

Chapter 4

Niles & Company

Unlike those of clergymen or politicians, the lives of mechanics are rarely given more than cursory notice, if that, in county histories. Their papers are seldom saved and their contributions almost never discussed in learned journals. The lives of the Niles brothers, iron founders and machinists, remain so obscure that not even their birthplaces or portraits can be found. (Virginia and Connecticut have both been said to be their original home, although the 1850 census records give Vermont.[138]) Only the most elementary reconstruction of their history is possible from the scant information presently available.

Whatever their origins, in 1834 Jonathan S. Niles (1804–1878) was a partner of George L. Hanks in the firm of Hanks & Niles, machinists, iron and bell founders.[139] Two years later his younger brother, James M. Niles (1809–1881), joined the firm, which was located in Cincinnati on the east side of Main Street between

---

[138] In his reminiscences (*Cincinnati Enquirer*, January 29, 1922) of early Cincinnati mechanics, Elijah L. Davis stated that the Niles were from Virginia. Connecticut is apparently a conjecture based on their retirement to Hartford. It was suggested in JOHN E. DIXON, *Lima-Hamilton* (New York: Newcomen Society [American Branch], 1948), p. 12.

[139] Cincinnati city directory, 1834. The earlier city directories of 1819, 1829, and 1831 did not list either J.S. or J.M. Niles.

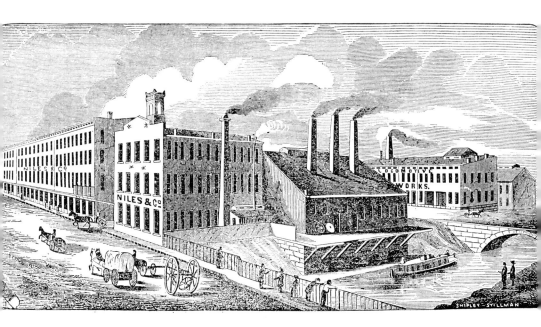

Figure 39.—VIEW OF NILES & Co. Note the locomotive works in the upper right corner. (From the 1853 *Cincinnati Business Directory*.)

Woodward and Franklin Streets, several squares north of the canal. By 1842 the firm had become Niles & Company and remained in its original location until 1845.[140]

At this time, 1845–1846, the Niles brothers bought out the sugar-mill manufacturing business of William Tift, whose shop was located on Third Street between Sycamore and Broadway. The manufacture of sugar mills was to become one of their most profitable lines of work, as it was for several other Cincinnati shops. In 1847 they were reported to have orders for 40 mills costing about $7,000 each,[141] in 1851 orders for 70 were on hand.[142]

---

[140] Hanks had withdrawn from the firm shortly before to form the new partnership of Hanks & McGraw, Bell Founders.

[141] *Cincinnati Commercial*, July 5, 1847.

[142] A handsome set of mechanical drawings for a Niles mill can be found in GUSTAVUS WIESSENBORN, *American Engineering* (New York, 1861) pl. 32. The Ford Museum has   Niles mill in its collection.

To accommodate their expanding business, which, in addition to sugar mills, included steamboat and stationary engines, machine tools, and rolling mills, the Niles brothers built a new machine shop on East Front Street at the drainage canal, formerly Deer Creek, convenient to the river and the shipyards at nearby Fulton. The *Cincinnati Gazette* of October 21, 1845, noted that the shop was nearly complete. The main building measured 40 by 105 feet; a foundry and blacksmith shop were on the first level and a machine shop on the upper stories. It was further reported that about 150 men would be regularly employed.

The Niles brothers did well in their new establishment and in the next few years became one of the largest and most affluent manufacturers in the city. Their prosperity prompted them to take on new and promising lines of work. Undoubtedly they observed the large business of their neighbor, Anthony Harkness, and, with many others, were stimulated by the growing demand for locomotive engines. As early as August 5, 1850, J. L. Whetstone wrote to G. E. Sellers, "Niles is going into the Locomotive business and I think Greenwood has strong notions the same way . . . . He says Askew is very keen to go into the locomotive business."[143]

Apparently no locomotive work was actually undertaken until almost a year later. The first mention of such activity is an item in the *Cincinnati Enquirer* of April 18, 1851, which states that the Madison and Indianapolis Railroad had contracted with Niles for two locomotives to be delivered in October. In the same article the *Enquirer* notes that "Messrs. Niles & Co. have recently added largely to their before extensive establishment" and indicates that the Madison and Indianapolis order was Niles' first contract for a locomotive: "To their *new* enterprise we certainly wish them every success" [italics supplied]. A letter dated August 31, 1851, from Coleman to his brother George Sellers, also indicates that Niles had taken up railroad work, for Coleman mentions that he had hoped to find a tender-truck pedestal pattern suitable for the Panama Railroad engines among those on hand at the Niles shop.

---

[143] Peale-Sellers papers. The *Cincinnati Gazette* for August 8, 1851, also notes that Greenwood was contemplating the manufacture of locomotives, but there is no evidence that he engaged in such work beyond the part he played in the construction of the Sellers engines for the Panama Railroad.

In the 1851 annual report of the Madison and Indianapolis Railroad, dated March 10, 1852, it was reported: "Messrs. Niles & Co. of Cincinnati, were not able to deliver two freight engines during the year. One of them has since been taken by the Dayton and Western road, and the other we expect to receive within a few days." However, a study of later reports of this company failed to reveal any Niles locomotives on its roster. A few Niles engines were reported on the Dayton and Western Railroad when it was leased by the Little Miami Railroad in 1865. Unfortunately, the date of their construction was not mentioned, so it is impossible to determine whether they were among the first locomotives built by Niles.

On August 8, 1851, the *Cincinnati Gazette* announced:

> On Congress Street, east of Butler, the Messrs. Niles have made a purchase of a large lot, near half a square, from Charles Fox, Esq., and have commenced the construction of a large foundry and machine shop (an addition to their already mammoth establishment on Front Street). This new building and yard is to be devoted, we are informed, to the manufacture of locomotives and cars for railroads.

With the completion of this new shop, the Niles brothers began to operate two businesses. The old shop on East Front Street became known as the Niles Works (later Niles Tool Works) and was devoted to the production of sugar mills, machine tools, and other products. The Niles Works was a joint stock company which was comprised, in addition to the Niles brothers, of the following partners: Leander H. Carey, Stephen Allen, S. H. Gilman, and Charles W. Smith. The new shop on Congress Street was under the direct proprietorship of James and Jonathan Niles; it was there that most, if not all, of the locomotive work was done.

By early 1852 Niles & Company was in full production of locomotives, a fact amply witnessed by the number of engines built. Fully ten Niles locomotives are listed in the annual reports of various roads for that year, and more were certainly built. Business was such that the proprietors were encouraged to increase locomotive production.

As previously mentioned, Coleman Sellers joined the firm in January 1852 after the failure of his brother's business. He was hired as an accountant and timekeeper. At that time Z. H. Mann, mentioned in connection with Anthony Harkness, was in charge

Figure 40.—COLEMAN SELLERS, superintendent of Niles & Co.'s locomotive department from 1853 to 1856.

of design, while N. E. Thorn was superintendent of the shop.[144] Coleman's work was not confined merely to keeping the books, for a diary which he kept with varying regularity between 1849 and 1854 records shopwork he performed as an assistant foreman. Some of these entries present a fine picture of the daily happenings at Niles & Company. This diary, which is to be found in the Peale-Sellers papers, is the basis for what follows on the next few pages.

On January 19, 1853:

Moved an engine from the shop to the Little Miami Railroad: (named Judge Gebhart) It was an engine built for the 4' 10'' gauge but altered to 4' 8½'' by moving the wheels in and moving the tire—After

[144] *American Railroad Journal* (December 3, 1853), vol. 9, p. 774. Here the superintendent is referred to as N. G. Thom.

we had got the engine out of the shop and turned round on the street one of the hand holes blew out and let all the water out of the boiler, which detained us a great while yet in spite of all delays by way [?] of hard work, and no delay at noon for I worked the men all noon time, we had the engine and tender on the track by 5:00 o'clock p.m.

Eight days later he remarked "moved the engine 'Clarke' to the Miami Road in the unprecedented short time of one ½ day."

On February 16, 1853, Coleman noted that he had been temporarily made foreman of the locomotive shop but was expected to handle his bookkeeping work as well. This proved a heavy burden, but Jonathan Niles, who managed the business while James spent most of his time on selling trips, put Coleman off with the advice that he must prove himself worthy of a permanent promotion. Sellers continued his dual role for another two months but became so weary of giving the Niles "two days of work for one of pay" that he was ready to quit. On April 20, 1853, he was made foreman, relieved of his accounting duties, given an annual salary of $1,200, and instructed to be at the shop for the usual business day of 7:00 a.m. to 5:15 p.m. While Coleman's diary and letters do not record a direct opinion of his employers, one gets the general impression that it was not an entirely warm or congenial relationship. The Niles brothers appear to have been preoccupied with business affairs, haunted by the ever-present possibility of failure, bankruptcy, and the other calamities which daily threatened to ruin them and their competitors in the free-market economy of the times. This is not to say that they were necessarily unfeeling men, driven by the compulsion of gain alone; yet nowhere does either Coleman or George Sellers speak warmly of their daily associates and employers.

It is interesting to note that one of the rare surviving comments on the personality of James Niles, which appeared in his obituary, states that he was a ". . . gentleman of genial and pleasing manners,"[145] while Coleman Sellers described him as quiet and retiring when at the shop or around business associates, lest he let slip some detail of their affairs to a rival.

The Niles were so satisfied with the profits and prospects of locomotive work that by June 1853 they were contemplating disposing

---

[145] *Hartford Daily Courant*, August 8, 1881.

of their sugar-mill manufacturing business, which had been to this time their major line. Not only was railway engine building lucrative, but the Niles were becoming discouraged with the southern trade, whose poor credit record now required them to demand a 50 percent down payment. This in turn discouraged the capital-poor Louisiana planter from purchasing Niles machinery.[146]

During 1853 the *American Railroad Journal* reported that the Niles were building two or three locomotives per month;[147] the next year they built over 40.[148] The *Railroad Advocate* gave a bright report on the Niles:

> The West is fast becoming competent for the supply of the equipments of her own roads. First in starting, and in its present extent, is the large Establishment of the Messrs. Niles & Co., of Cincinnati; which of late years has been extensively devoted to the manufacture of Locomotives and Railroad Machinery. On a visit to Cincinnati, we were not prepared to find so large and so well arranged and well equipped works as theirs, engaged in the business. We owe an apology to the West for such a confession, but we had, until then, supposed that nearly all Western roads were dependent on the East for Locomotives. We learned our mistake, however, by a little observation upon some of the leading roads of Ohio. We found, not only upon these roads, but in the shops of the Messrs. Niles & Co., Engines of the best construction and capabilities.
>
> Being a little critical in such matters, we took occasion to observe the strength and permanence of the cylinder-fastenings of these engines, the strong and durable framing employed, and the excellent fitting-up of the link-motions. All the joints about the links and valve gear are made with provision for being kept up in close and accurate adjustment, and are well case-hardened to provide against wearing. The materials, procured in Cincinnati, are also of the best kind. Altogether, it is creditable to the West to possess so complete and extensive works devoted to a most important branch of industry, and it is creditable to a greater degree that these works have attained to an honorable distinction for the excellence of their productions.[149]

---

[146] G. E. Sellers to Horatio Allen, June 15, 1853, Peale-Sellers papers. Sellers suggested that Allen might be interested in taking over the sugar mill line for production at the Novelty Iron Works. Apparently he was not interested.

[147] (October 15, 1853), vol. 9, p. 666.

[148] *Railroad Record* (March 1855), vol. 3, p. 86.

[149] (November 18, 1854), vol. 1, p. 1.

Oddly enough, at what was apparently the time of its greatest activity, Niles & Company began to reduce their production of locomotives. On March 25, 1854, Coleman Sellers noted in his journal: "Mr. Niles says [he] has made up his mind to cut down the force on Locomotive work and only turn out an engine every two weeks. I am very sorry for it, for I have just got the shop organized to turn out one a week with care." Since railroads, being speculative, were unusually sensitive to any financial disturbances, a short-lived panic caused by a ". . . depression in the monetary affairs of the country . . ." [150] induced the firm to reduce its shop force and inventory. After this time Niles continued to show varying moods of caution and optimism regarding locomotives. One important factor in the firm's indecision was the conflict of the brothers' opinions on the soundness of pressing further into the increasingly competitive and specialized field of locomotive building.

The *Railroad Record* reported that because the financial upset had caused many railroads to cancel orders for new equipment and so many shops were entering this line of work, locomotive production was far greater than the existing demand.[151] They reassuringly commented that both Niles and the Cincinnati Locomotive Works were not lacking work. This report, as we have seen, was too optimistic, for business did not pick up fully in that year or the next or, in fact, at all

In a letter (December 23, 1855) to his mother, Coleman Sellers very neatly summarized the principal difficulties of the trans-Appalachian locomotive builders:

> Locomotive building seems to be a poor business in the West—our market is limited to the Western roads, which are all very poor and do not at present seem inclined to find cash to pay for engines when they can go east and buy cheaper machines (which nine cases out of ten are made to sell and not to run) and pay for them in bonds which bonds can be negotiated in the east to much better advantage than they can here—Mr. James Niles seems determined to drive his brother in to giving up the business. It is probable that unless things change wonderfully they will make some change in their business and sell out their shop.

The Niles brothers gave constant consideration to closing out the

---

[150] *Railroad Record* (October 19, 1854), vol. 2, p. 565.
[151] Ibid.

Figure 41.—JOHN L. WHETSTONE, 1821–1902, chief designer of Niles & Co., the most gifted and original engineer employed by any of the Cincinnati locomotive builders. This photograph was taken in 1868. (*Photo courtesy Historical and Philosophical Society of Ohio.*)

production of locomotives from this time forward, until the great Panic of 1857 made the final decision for them.

Coleman apparently felt that since his employers were so skeptical of the future of the locomotive business, and since he was superintendent of that particular department, his prospects for advancement were poor. He began to look for employment elsewhere. In January 1856 he accepted a position as an engineer with his cousin, William Sellers, who operated the well-known machinery company of that name in Philadelphia. Within a few years he was a member of the firm and had begun his steady rise to fame as a nationally recognized engineer, first as president of the Franklin Institute, then as president of the American Society of Mechanical Engineers. Undoubtedly, the high point of his career was his position as chief engineer of the Niagara Falls Power Company.

Many years later he wrote to Angus Sinclair of his earlier experiences with Niles & Company. This sketch, printed in Sinclair's *Development of the Locomotive Engine*, presents a brief but accurate outline of that concern.[152] (A drawing of a small 4–2–0 named *Cincinnati*, reproduced in Sinclair along with Sellers' narrative, is labeled "Early Niles Locomotive," but it was actually a machine

---

[152] Pp. 360–362.

built by the Vulcan Foundry [Tayleur & Company, Warrington, England] in 1835 for the South Carolina Railroad.)

While the Niles brothers brooded over their ever-fluctuating fortunes, some of the most unusual and mechanically advanced locomotives built in the Midwest were under construction in their extensive factory. The genius of this work was John La Fayette Whetstone (1821–1902), a son of the well-known Cincinnati pioneer of the same name. His early life strangely parallels that of his later business associate George E. Sellers (see p. 47). He had a marked inclination toward the fine arts and as a young man was known to be a sculptor and artist of some ability. The elder Whetstone guided his son toward mechanics as a more suitable career, and Whetstone adopted it with enthusiasm.

The earliest demonstration of Whetstone's mechanical imagination was his development of the radial valve gear used on Sellers' grade-climbing locomotives built for the Panama Railroad in 1851 and 1852. The use of radial valve gears, which impart a constant lead to the valve by means of the crossheads motion, results in a more precise and economical distribution of steam to the cylinders. Although valve gears of this type were widely used in Europe from the late 1840's on, they received scant attention in this country until 1904, when the motion was reintroduced in this country by the spectacular performance of the mallet number 2400 of the Baltimore and Ohio Railroad.[153] As early as 1838 Seth Boyden built a locomotive using a form of radial valve gear, but, unfortunately, little data is available on it.

Practically no mention of Whetstone's introduction of radial valve gear is to be found other than a few notes in Sinclair's book and in the *Journal of the Franklin Institute*. While Sinclair cites Whetstone's motion as the earliest American form of this device, he incorrectly states that it was used on engines built by Niles in the early 1840's—at least ten years before Niles was in the locomotive

---

[153] William Mason used a radial motion of the Walschaert pattern on his patented double truck locomotives as early as 1874–1875. Herbert Fisher, in *Railway and Locomotive Engineering* (July 1909, vol. 22, p. 302), speculated that Mason may have been inspired to experiment with radial motions after seeing the *Defiance* and the *Champion* in operation on the Lehigh Valley Railroad. Built by Niles & Company, these engines were equipped with Whetstone's valve gear (see Chapter 3, pp. 84–88).

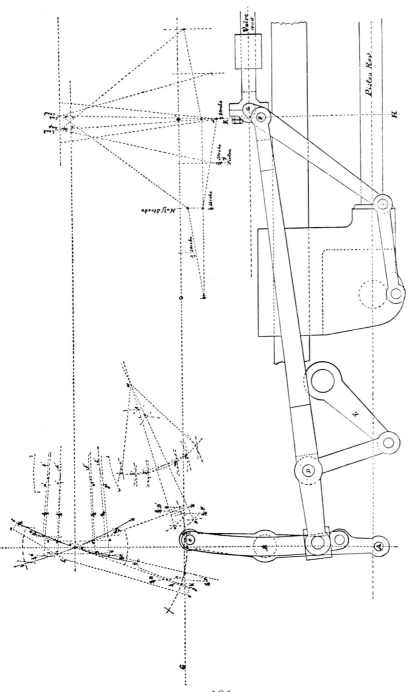

Figure 42.—Whetstone's Radial Valve Gear was developed for use on the Sellers grade-climbing engines in 1851. It was subsequently used on a number of locomotives built by Niles. Note the solid link which is attached to the rocker. The eccentric straps and driving axles are not shown. (From the *Journal of the Franklin Institute*, Feb. 1882.)

101

business. Sinclair erroneously states that Whetstone's valve gear used "a small return crank" (an outside-mounted eccentric crank), when actually a single eccentric for each cylinder, mounted as in the usual Stephenson motion on the forward driving axle inside the frame, drove the valve in combination with the crosshead.[154] Many years later, at the suggestion of Coleman Sellers, Whetstone prepared a precise mechanical description of this valve gear which appeared in the *Journal of the Franklin Institute*.[155] (The following account is based on this article and pertinent extracts are reprinted as Appendix 3.)

Whetstone developed his form of the radial valve gear while working with the Sellers on the grade-climbing locomotives for the Panama Railroad in 1851 and 1852. Because the highly complex grade-climbing machinery entirely occupied the space between the locomotive frames, preventing the usual placement of four eccentrics on the front driving axle, a conventional hook or link motion was impossible. Therefore Whetstone devised a valve gear which required only two eccentrics on the forward driving axle; all other parts were mounted outside the frame.

This motion was used on the two large Sellers grade-climbing engines (see p. 84) built for the Coal Run Improvement Railroad and on many other locomotives built by Niles & Company. To my knowledge, however, it was not employed by any other builders. In April 1854 Whetstone also devised a practical system for the layout of Stephenson link motions so that the admission and cutoff were absolutely uniform in every position of the link, whether forward or in reverse.[156] This was accomplished by preparing a series of full-size layout drawings and reapportioning each part of the mechanism until it would cut off square at all points. This graphic layout was further checked by the construction of full-size, adjustable working models. This method was followed for each class or new design of engine built and, when completed, the layout and model could be followed for any other engines built on the same plans. With this method tedious and makeshift adjustments or alterations on engines being fitted up in the shop could be avoided,

---

[154] *Development of the Locomotive Engine*, pp. 362, 457–458.

[155] (February 1882), vol. 113, pp. 106–117.

[156] Coleman Sellers' journal, April 13, 1854, Peale-Sellers papers.

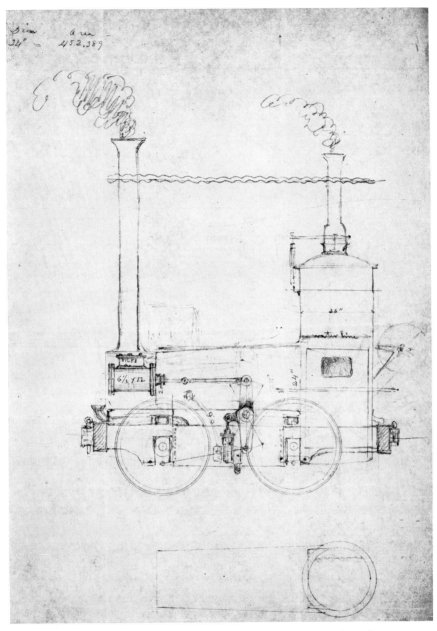

Figure 43.—A STUDY SKETCH for a compensating lever locomotive probably prepared by Coleman Sellers about 1855. Note "Niles" marked on the valve box. From the Peale-Sellers papers. (*Courtesy American Philosophical Society.*)

103

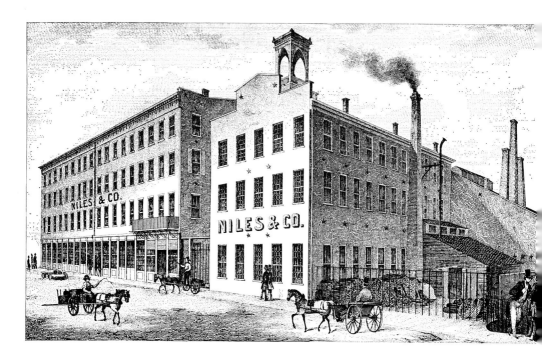

Figure 44.—Niles & Co. Shops at East Front Street and Deer Creek, 1851. Note railing of the Deer Creek bridge at extreme right. This plant later became known as Niles Works; locomotive work was done at the new factory on Second Street. (From Charles Cist's *Cincinnati in 1851.*)

since they had already been made in the drafting room.

The *Railroad Advocate*, in its issue of January 12, 1856, congratulated the inventor: "The principle is beautiful, and the locomotive builders of this country could well afford to unite in a handsome reward to Mr. Whetstone, for the simple process by which he secures the results [uniform steam admission] to which so many unsuccessful efforts have been directed." While describing the details of his geometric method of valve-gear layout, Whetstone readily admitted, however, that the otherwise commendable precision of his motion produced a ". . . uniformity of lead . . . not favorable to the attainment of the highest speeds required in passenger traffic."[157]

---

[157] See footnote 155.

It was not until Whetstone succeeded Z. H. Mann as chief designer, according to Coleman Sellers, that the products of Niles reflected any important improvements.[158] Sellers claimed that Whetstone experimented with a plate frame but gave it up for the more familiar and flexible bar frame. This so-called plate frame was undoubtedly either a slab bar frame or a composite type, often called a riveted frame, built up of two rails about 5 inches wide and ½ inch thick between which the pedestals were fastened. There is some evidence, as will soon be pointed out, that a true plate frame, so common in the British locomotive, was used.

Although it is possible that Whetstone was inspired by the Camel engines built by Winans, there is reason to believe that some true slab frames were used on Niles engines. The slab frame was fabricated in the same manner as the conventional bar frame, but a single, thin, deep-throated bar was used in place of top and bottom rails to connect the pedestals. However, slab frames were little used in this country except by Breeze, Kneeland & Company of the New York Locomotive Works, Jersey City, New Jersey. Evidence that Niles used this type of frame is found in Coleman Sellers' journal for February 4, 1854, in which he speaks of the locomotive *Mercury* of the Springfield, Mount Vernon, and Pittsburg Railroad as having a "slab rail frame."

More evidence of Niles & Company's interest in plate frames is a rough sketch (fig. 43) of a curious 4-wheel locomotive included in the papers of Coleman Sellers at the American Philosophical Society. The only identification other than dimensions is the word "Niles" on the side of the valve chest. It is probably a study drawing inspired by the compensating lever engine devised by Alba F. Smith of the Cumberland Valley Railroad. A lengthy report on one of these machines, built by Seth Wilmarth at his Union Works, Boston, Massachusetts, for that road, appeared in the *Railroad Advocate* on October 27, 1855. Judging from the number of comments on Niles & Company found in its columns, it is not unlikely that Sellers and Whetstone were regular readers of the *Advocate* and were prompted by Wilmarth's report to prepare the sketch shown here. Notice that the feedwater pump, attached to the

---

[158] SINCLAIR, *Development of the Locomotive Engine*, p. 361.

Figure 45.—THE MARIETTA AND CINCINNATI RAILROAD'S
Vinton, built by Niles in November 1855. Shown here at
Loveland, Ohio, in 1872. (*Photo courtesy Ohio State Historical
Society.*)

frame between the driving wheels, is driven by an arm forged onto
the left side of the compensating lever. Several interesting features
are illustrated by the "T" boiler, which has what appears to be a
side fire door: the use of India-rubber blocks for springs and the
suggestion of a winch or possibly a hose reel on the back of the boiler.
It is a curious and tiny machine, undoubtedly intended for industrial service. The significance of this improbable locomotive
is to show that Sellers and Whetstone did in fact consider the plate
frame. If used by them to any extent it constituted a distinct
departure from regular American practices.

Whetstone's attraction to full-size layouts was carried over into
the shop itself, where a mammoth T square was used to determine
center lines and otherwise ensure accurate and precise assembly of
the locomotives. Although little else is known of Whetstone's
other reforms in design and shop practices at Niles, he did obtain
two patents for improvements in locomotive engines. One of these
(27850, April 10, 1860) concerned itself with a 2-wheel guiding
truck for 6-wheel engines. This device, although apparently not
adopted or put to a practical test, anticipated to a great extent the
Hudson-Bissell truck of 1864, which permitted the introduction of

the highly successful Mogul and Consolidation types of locomotives.[159] The other patent (33760, November 19, 1861) concerned a modification of Sellers' grade-climbing locomotive in which the center rail was eliminated and the adhesion wheels gripped one of the running rails. The proposed engine had two sets of gripper wheels, employed an extraordinary number of gears, and, considering the tremendous overhang of the machinery on one side of the engine, was a doubtful improvement on Sellers' patent of 1850.[160] In fact, Sellers had dismissed this idea as impractical many years earlier.

Whetstone's mechanical ability was by no means always praiseworthy: many of his ideas were impractical, and his preference for full graphic layouts over mathematical calculation of mechanical problems suggests an innocence of analytical competence, even though the practical values of full-size layouts are still recognized today. But Whetstone's innovations are of importance in dispelling the misconception that midwestern builders were unimaginative imitators of established eastern builders. While the Cincinnati builders, notably Niles, showed some originality, they did have the good sense to follow the better practices of the more advanced eastern builders. Particularly influential were the shops located at Paterson, New Jersey, such as Rogers and the New Jersey Locomotive and Machine Works, whose use of wagon-top boilers, widespread trucks, outside connections, etc., was favored by the Cincinnati builders. Certain advances in the design of the Cincinnati locomotives can be traced to the drafting boards of such gifted eastern engineers as Walter McQueen, W. S. Hudson, and William Mason, to name but a few. Also, several superintendents of midwestern shops were trained in the East: James Tull of Palm & Robertson apprenticed at Norris, and James Water of Menominee Foundry was from Portland.

Niles built a large number of outside-connected 4-4-0's in 1853 and 1854 for the new broad-gauge Ohio and Mississippi Railroad.

---

[159] For a more complete discussion see John H. White, "Introduction of the Locomotive Safety Truck" (paper 24 in *Contributions from the Museum of History and Technology: Papers 19-30*, U.S. National Museum Bulletin 228; Washington: Smithsonian Institution, 1963), pp. 117-131.

[160] A picture of the patent model for this invention is in *Scientific American* (August 11, 1906), vol. 95, p. 101. The present location of the model is unknown.

Figure 46.—THE OHIO AND MISSISSIPPI RAILROAD'S No. 19, built by Niles in October 1854. Shown in 1870 as originally built for 6-foot gauge. (*Photo courtesy L. W. Sagle.*)

Figure 47.—A NILES-BUILT OHIO AND MISSISSIPPI LOCOMOTIVE, probably constructed in 1854. Shown here as rebuilt for standard gauge in about 1871. Original number and scene unknown.

The *Railroad Record* reported, "Yesterday was shipped on board a flatboat, to be shipped to Lawrenceburgh, for the Ohio and Mississippi Railroad, one of four large engines being constructed here for that road. This is the first wide gauge, six foot gauge engine built west, and has its cylinders parallel with the shafts.[161] Niles is known to have built at least twenty locomotives for this road. American types, they were heavy for the period, but much of the weight is explained by the larger machinery necessary for the wide gauge. Their cylinders, for instance, were 15 inches in diameter with a 20-inch stroke and generally required two days to bore. The work was done with a horizontal boring bar and a 36-inch lathe. The same method was used by Baldwin and most other builders, since it apparently was the best and quickest method known at that time.[162]

Two remarkably detailed descriptions of Niles locomotives appeared in the *Railroad Advocate* in 1856. The engines are described as handsome and built on a progressive design. The components are stated to be laid out in a tasteful, thoughtful manner. The boiler, set low, had a high wagon top for ample steam room. Unusually large steam ports 1 inch wide by 18 inches long were used for 15 x 22 inch cylinders. The disadvantages associated with large ports were overcome by balanced valves. The light cylinder fastening made without a conventional saddle was praised. Both articles are reprinted as Appendix 4 for the reader wishing a more complete mechanical description.

Why hadn't Niles or, for that matter, the other western builders captured the local market for locomotives? They built an honest, progressive machine, employed alert and competent technicians and designers, and could save local railroads the shipping charges, generally about $800, from eastern shops. Although many other factors were involved, two basic reasons were prejudice against the western shops and the easy credit of their eastern adversaries. Most of the western railroads were built with eastern capital and were staffed by men from the same section. They not unnaturally preferred the

---

[161] (December 8, 1853), vol. 1, p. 643.
[162] *Journal of the Franklin Institute* (January 1900), vol. 149, p. 10.

Figure 48.—The "Sidney," built by Niles for the Bellefontaine and Indiana Railroad in July 1853. Shown here as rebuilt in 1856 at the Galion, Ohio, shops of the Cleveland, Columbus, Cincinnati, and Indianapolis Railroad. Scene is Brightwood, Ind., in the 1870's. (*Photo courtesy Ohio State Historical Society.*)

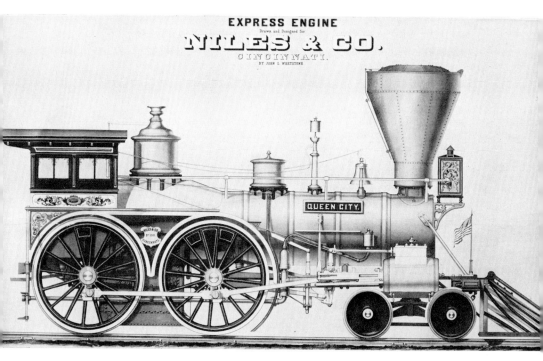

Figure 49.—The handsome symmetry of the *Queen City* is a tribute to Whetstone's skill as a designer. While the road for which this machine was built cannot be determined, the *Railroad Record*, on July 10, 1856, noted the publication of this lithograph.

Figure 50.—BUSINESS CARD OF NILES & CO., about 1855.

machines with which they were most familiar, resorting to western-built machines only if in immediate need of additional power. Just as naturally, the eastern banker preferred to see orders for locomotives stay in New England, because he either had an interest in one of the locomotive works or simply desired to promote the local economy. The *Railroad Record* summarized the problem:

> the majority of engineers and superintendents upon our western roads, are Eastern men, and come amongst us with Eastern prejudices, being unable to comprehend how machinery can be manufactured in a country where, but a short time since, the Indian roamed untrammelled in his native forests.[163]

Some months later, the same journal commented more heatedly:

> Objection has been made to our Western locomotives—mostly by "Eastern men". . . that they have not the *finish* of the Eastern shop work. This *may* have been the case heretofore, as it has been the aim of our mechanics to put the greater portion of the price of their work *into* rather than *upon* their locomotives, that while the Eastern gingerbread work is half the time in the repair shops theirs may be doing its duty upon the road.[164]

---

[163] (May 4, 1854), vol. 2, p. 148.
[164] Ibid. (October 19, 1854), vol. 2, p. 565.

111

Figure 51.—The "Elkhorn" of the Chicago and Northwestern Rail-road, built by Niles in 1858. Scene is Askew Arch Culvert, near Rockford, Ill. (*Photo courtesy A. W. Johnson.*)

(A report favorable to Niles & Company and its locomotives appeared in the *Railroad Advocate* of June 7, 1856, and to a degree substantiates the paternalistic defense of the *Record* just cited. It is reprinted herein as Appendix 5.)

While some western railroads encouraged local machinists to manufacture locomotives, as the Little Miami did Anthony Harkness, most lines felt called upon to justify such purchases. This was undoubtedly done to reassure the stockholders that money was not being spent carelessly on inferior machinery. The Marietta and Cincinnati Railroad, for example, commented in its sixth annual report (1856):

> Five additional locomotives are now nearly completed at the works of Messrs. Niles & Co., of Cincinnati, who have built the ten superior engines purchased the present year and now in use on the road. The power and speed of these engines, the economy with which they may be run, and their other excellent qualities make them fully equal to the

Figure 52.—The "Sabine," Built in 1857 for the New Orleans, Opelousas, and Great Western Railroad, was the last Cincinnati-built locomotive in existence. It was scrapped in 1942. (*Photo courtesy Southern Pacific Railroad.*)

best locomotives manufactured in the most celebrated Eastern establishments.

In a similar vein the advertisements of Niles & Company and Moore & Richardson stated, almost apologetically, that their products were equal to those of the best eastern shops.

Credit was the most potent reason for the eastern domination of the locomotive market. With greater resources than most western shops could muster, many eastern builders would extend long-term sales contracts or accept shares of stock, usually at a discount, for a

113

Figure 53.—The "Vincennes," Built by Niles in 1855 or 1856 for the Evansville and Crawfordsville Railroad.

large portion of the selling price. The Lawrence Machine Shop is known to have accepted land from railroads in Indiana, Illinois. and Nebraska in return for locomotives.[165] Undoubtedly other eastern shops of no less means were willing to speculate on the rising value of western land. Conversely, the railroads were apparently delighted to accept cheap locomotives on easy terms, since they could then invest their cash in land or extensions of their lines in the belief that rapid settlement of the country would increase land value and traffic so that new and better equipment could be purchased at some future date. It was this shortsighted optimism, coupled with no small degree of opportunism, which contributed to the huge over extension of credit (mainly to booming midwestern rail lines) that resulted in the Panic of 1857.

While commenting on the state of the locomotive building industry and the disastrous effects of the financial collapse in 1857, the *Railroad Advocate* stated as a matter of history: ". . . locomotives

[165] *American Railway Times* (February 8, 1862), vol. 14, p. 46.

were no longer bought in large quantities with cash, but on long credits and with in some cases precarious securities. This kind of pay rendered it impossible for any but houses of ample capital or of established reputation to succeed."[166] The Niles brothers did not have the capital, nor were they willing or able to accept large quantities of railroad securities any more readily than when they found that credits extended to customers in the southern sugar-mill trade were not always honored. By June 1857 it was reported that Niles, finding the construction of riverboat engines more attractive, was no longer interested in promoting locomotive work.[167] Although no definite reasons for the closing of the Niles shops have been discovered, enough data exist to determine the date with reasonable accuracy. The last advertisement for Niles & Company that appeared in the *Railroad Record* (June 25, 1857) indicated a lack of interest in soliciting new orders for locomotives. The plant apparently remained open, for the *Cincinnati Commercial*, discussing a court case in its issue of May 11, 1858,[168] said that the business was controlled by two proprietors, J. S. and J. M. Niles Sometime between this date and June 1859 the Niles brothers retired, withdrawing from active management of both foundry properties. The Congress Street shop was taken over by the I. & E. Greenwald Company as an addition to its machinery business.[169] The shop on East Front Street continued to operate, retaining the name of Niles Works, with H. A. Jones as president.[170] Under the management of George Gray, James Gaff, and Alexander Gordon, Niles Works gradually entered the field of machine tools, and within a few years this became its major line of work. In about 1872 the plant moved to Hamilton, Ohio, but continued to use Niles as the corporation title. Today a division of the Baldwin–Lima–Hamilton Corporation builds machine tools under the name Niles, in respect to the founders of the long defunct Niles & Company.

The Niles brothers left Cincinnati sometime in 1859 (they are

---

[166] (June 6, 1857), vol. 4, p. 4.

[167] Ibid.

[168] The article itself, concerned with a suit between the city and Niles & Company over the collapse of a portion of the Deer Creek Culvert, is not otherwise relevant.

[169] CHARLES CIST, *Sketches and Statistics of Cincinnati in 1859*, p. 285.

[170] Ibid., p. 289.

not listed in the 1860 Cincinnati City Directory) and moved to Hartford, Connecticut, where they each built large homes and retired in very comfortable circumstances, only on rare occasions taking an active part in local business. Jonathan lived until March 1, 1878, his brother James until August 8, 1881.[171]

A fine example of Niles & Company locomotives survived as late as 1942. This was the *Sabine*, built in 1857 for the New Orleans, Opelousas, and Great Western Railroad, later part of the Southern Pacific system. In 1897 the engine was sold to a sugar refinery in New Iberia, Louisiana, for use as a switcher. After standing idle for some years, the *Sabine* was repurchased in 1923 by the Southern Pacific for display at Lafayette, Louisiana, where it quietly reposed for 19 years as a monument not only to early railroading but also to the long-silent Cincinnati locomotive shops. On October 5, 1942, the *Sabine* was junked in a public ceremony to promote popular support of a local wartime scrap drive.

---

[171] From obituaries of the brothers (copies supplied by the Connecticut Historical Society). These accounts make no mention of either brother's birthplace or reason for retiring to Hartford.

Figure 54.—Advertisements of Niles & Company appearing in railroad journals during the 1850's.

LOCOMOTIVE WORKS.

NILES & CO.,
CONGRESS STREET, CINCINNATI,

BUILD to order Locomotives of any required size or plan, and are prepared to execute all orders in their line with promptness.

Orders solicited for iron and Brass Castings, Flue and Cylinder Boilers, Tyres, Tyre Lathes, Planing Machines, and other tools, Shafting, &c . &c.
feb. 13 1855 6m.

NILES & CO.,
CINCINNATI, OHIO,

MANUFACTURE TO ORDER

Locomotive Engines & Tenders,
Of any arrangement or weight required, and which, for Economy, Durability, and Efficiency, are offered in comparison with any work produced in the country.

ALSO, GENERAL

MACHINE AND FOUNDRY WORK.
Cincinnati, Nov. 18, 1854.

Chapter 5

The Covington

Locomotive Works

Covington, Kentucky, located just across the Ohio River from Cincinnati, is so closely associated with the larger city that an account of the Covington Locomotive Works is a natural adjunct to this study, even though the firm was short-lived and is known to have produced only six locomotives. Its history is obscure and only the barest evidence attests to its existence.

Through the agency of A. L. Greer, a Covington capitalist, the affairs of the Covington Works were closely associated with the Covington and Lexington Railroad. Greer was involved with that line in 1852 as agent in charge of purchasing rails and was for several years a member of the Board of Directors.[172] It was undoubtedly this association that prompted him to organize the Covington Locomotive Works in 1852–53. The only orders it is known to have received were from the Covington and Lexington.

The earliest notice of the works appeared in the *Covington Journal* of April 16, 1853. It stated that the shop buildings were completed and that the machine tools were being installed. The works was located on both sides of Philadelphia Street between Third Street and the Ohio River. The foundry, a massive stone structure with a fireproof roof, was 100 feet square and 40 feet high. It was described as airy and well lighted. The 4-storied machine shop, of similar construction, was 160 feet long and 45 feet wide; the erecting

---

[172] *Annual Report of the Covington and Lexington Railroad*, 1852; *American Railway Times* (January 12, 1854), vol. 6, p. 16.

shop was 80 feet square, the blacksmith shop 250 by 45 feet, and the boiler shop 200 by 60 feet. A large yard and several smaller shops for brass founding and tin work completed the establishment. It was certainly a plant equal in size to any works west of the Alleghenies. The last of several brief comments in the *American Railroad Journal* concerning the opening of the Covington Works appeared in the December 10, 1853, issue (p. 791). It stated that the works would soon be in full operation.

The Covington and Lexington Railroad gave Greer an order for four locomotives. One was reported near completion in November 1853, but was not delivered until May of the next year. On February 25, 1854, the *Covington Journal* stated that the works was still not fully equipped or in full operation. Nevertheless, it was claimed that orders for ten locomotives were in hand. Four, as already mentioned, were destined for the Covington and Lexington, two for the Little Miami, and the other four for an unnamed road in Indiana. No record has been found that any engines were delivered other than the four for the Covington and Lexington. It is probable that the financial panic of 1854 caused the other roads to cancel their orders or that they were discouraged by Greer's slow delivery and transferred their orders to another builder. It is also possible that these six engines, if built, were sold to other buyers.

Greer had hardly started to build locomotives when it was announced that he was opening a car plant.[173] Several months earlier the Hart & Dryer Car Works of Covington had been reported in operation. Greer also had to face competition from the Fulton Car Works, established in Cincinnati several years before. Little is known of Greer's car-building activities, but it can be assumed that with the several builders of good reputation in the area his prospects were not good.

The first locomotive to be completed, the *Covington*, was many months in the building. An interesting account of its first trip appeared in the *Railroad Record* of May 4, 1854 (p. 146):

"COVINGTON" LOCOMOTIVE—TRIP TO "BOYD'S."

It is known to our readers that the Covington and Lexington Railroad is now opened 55 miles—within 5 miles of Cynthiana, the county-

---

[173] *Covington Journal*, February 25, 1854.

seat of Harrison. At the invitation of the proprietors of the Covington Locomotive Factory, we took, on Saturday, a trip to "Boyd's," 50 miles. The occasion of the party was the first trip of the "Covington,"—the first locomotive built in Covington.

The Covington Locomotive Works have just got into operation, and are on a very large scale, calculated to do an immense business. They have gone on so quietly that few we believe, know really what has been done; most persons will be surprised by the result accomplished. The Covington Locomotive Company consists, we believe, of A. L. Greer, of Covington; D. A. Powell, of this city; Mr. Fagin [Feger], [174] machinist for the Reading R.R. Mr. Finch, and two other gentlemen, making six partners with a capital of $500,000. The works are built on the most extensive scale, the principal building being of stone, and 200 feet in length. They now employ 130 hands, and when in complete operation, can employ 700, which is equivalent to the employment and support of 3,500 persons. Mr. Fagin [Feger] is the superintendent, and is an accomplished machinist. When under full headway, this establishment will get out a locomotive each week, having ten constantly on hand in construction.

The "Covington" whose maiden work was begun on Saturday, is a full certificate, both in work and appearance, to the skill and success of the Covington Factory. It is a fine, strong, handsome engine.

Three more locomotives of the same pattern, the *Cynthiana*, the *Paris*, and the *Lexington*,[175] were built during the summer of 1854. The *Lexington* was delivered in August 1854; it is the last locomotive built by Greer for which any record can be found. However, Feger ordered boiler tubes from the Baldwin Locomotive Works in September 1854 and asked Baldwin's advice on setting tubes.[176] This clearly demonstrates that Greer was still in business and apparently had other locomotives under construction.

---

[174] Daniel H. Feger, draftsman and designer, later master mechanic of the Memphis and Charleston Railroad, had previously been with the Philadelphia and Reading Railroad (*American Railroad Journal*, December 1853, vol. 9, p. 74).

[175] Reporting in the *Railroad Advocate* (May 3, 1856, vol. 3, p. 1) on the consumption of valve oil, T. D. Davis, master mechanic of the Covington and Lexington Railroad, stated that the *Lexington* made the poorest showing of their locomotives, consuming an average of one pint of oil for every 15 miles. Only three Covington-built engines were mentioned, indicating that one had been sold or discarded after less than two years of service.

[176] M. W. Baldwin papers (MSS., Historical Society of Pennsylvania, Philadelphia).

119

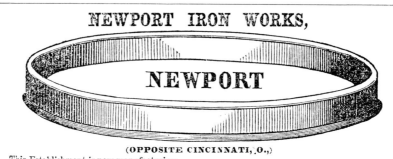

Figure 55.—ADVERTISEMENT APPEARING IN the *Railroad Advocate* throughout 1857.

Although the exact date cannot be determined, the works closed during the next few months, undoubtedly a victim of the hard times. A brief note in the *Covington Journal*, April 14, 1855, commenting on the depressed business conditions, stated that negotiations which might reopen the works were in progress. On May 19, 1855, the same paper announced the reopening of the plant by Cowles, Sickles & Company. D. H. Feger had been retained as superintendent. The new proprietors hoped to introduce oscillating-cylinder steam engines for riverboats (a form of engine popular in England but never favored in this country). [177] Intent on this venture, Cowles, Sickles & Company was not interested in further railroad work. The *Railroad Advocate*, October 20, 1855, stated that the Covington Works "are not, we believe, building locomotives at present." In fact, there is good reason to believe that the plant was out of business altogether by October 1855. The *Cincinnati Gazette* of December 28, 1855, announcing the sale of the Covington Locomotive Works, mentioned the suit of Snead, Collard, and Hughes against the firm, dated October 1855. Assuming them

[177] *Cincinnati Enquirer*, January 1, 1922.

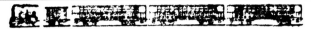

KENTUCKY
Locomotve and Machine Works
Covington, Ky.
WOLFF, SCOTT & FINCK,
MANUFACTURERS OF
Locomotives & all kinds of Railroad Rolling Stock.
ALSO
Stationary & Marine Engines.
ALL KINDS OF BOILER AND FOUNDRY WORK.
mch 28 '57 ly.

Figure 56.—ADVERTISEMENT APPEARING IN the *Covington Journal* of March 20, 1858.

to be creditors, it may have been that the suit was for foreclosure, since the assets of the works, including the property, machinery, and unfinished work, were to be sold at auction on January 19, 1856.

Earlier, on December 22, 1855, the *Covington Journal* had stated that A. D. Powell of Cincinnati, one of Greer's former partners, had purchased the plant for $100,000. This transaction failed to materialize, however, and the public auction remained set for January 1856. The results of the auction are in question since in the only known files of Covington papers the issues for January 20–25 are missing. It is safe to assume that the plant remained idle for over a year.

Daniel Wolff and several associates purchased the plant, according to a report in the *Railroad Advocate* of February 14, 1857. Wolff operated the prosperous Newport Rolling Mill and as early as 1853 had manufactured locomotive tires for Niles and Moore & Richardson.[178] The new proprietors, Wolff, Scott, and Finck, were to reopen the "long closed" works soon and were presently putting

[178] *American Railroad Journal* (October 15, 1853), vol. 9, p. 666.

121

the plant in good order. [179]   J. C. Finck was to be the superintendent. He was reported to have been with Moore & Richardson in the same capacity.

Wolff called the company the Kentucky Locomotive and Machine Works (not to be confused with the Kentucky Locomotive Works of Louisville, which operated a few years earlier). Two locomotives were built for the Covington and Lexington Railroad in June and September 1857. An advertisement for the Works was run as late as December 4, 1858, in the *Covington Journal*, but in all probability the plant closed at about this time, unable to survive the Panic of 1857 or eclipse the reputation of its rivals across the river.

---

[179] *Railroad Advocate* (February 14 and 28, 1857), vol. 3.

122

Appendix

Appendix 1

The shops and locomotives of the Cincinnati Locomotive Works (Moore & Richardson) are described in these two 1856 accounts from the *Railroad Advocate*.

CINCINNATI LOCOMOTIVE WORKS

THE works of Messrs. Moore & Richardson, known under the above title, have been devoted to the manufacture of locomotives since 1846, in which year they turned out the first one built in Ohio. The number of these engines built at the works since that time is over 120. In amount of business, and in the success of its manufactures, this establishment stands A 1. It is well enough to state some particulars of its extent and capacity, after which we may add something touching the details and performance of the locomotives built here.

The buildings and yard room occupy about a city square. The shops are somewhat detached, so that to follow out the feet and inches of every wall would be a little tedious. It is enough to say that the premises can comfortably accommodate 300 men in full employment.

The foundry is very large, being designed for casting heavy jobs of steamboat and rolling mill work, if required. It has two large air furnaces and two cupolas,—capable together of melting 20 tons daily. The cranes are very heavy,—the heaviest lift they ever had was six tons, but they can carry a greater load than that. The iron used for wheels and loco-motive work is mostly Hanging Rock. The truck wheels cast from this iron are very strong; we saw some which it had been found impossible for one man to break with a heavy sledge. The chill is about ⅜ inch deep and its line of mixture with the soft iron is very sharply defined,— that is, the chill does not extend back into the wheel in feathery or brush like veins. It is a question as to which chill is the best. The wheels cast here are cooled by Murphy's Patent Process, as lately described in the ADVOCATE.

The boiler shop is large and well supplied with tools. Cincinnati iron and Pittsburgh copper are used,—the latter for tubes.

The blacksmith's shop is large also. It has large forge fires and a heating furnace. A Nasmyth steam hammer of 3500 pounds is used for drawing out *tire*, frames, and axles. It deserves to be known that Moore & Richardson have been and still are large manufacturers of American *tires*. They have used but four sets of imported tires on all the engines they have built, having made the others in their own works. From what we can learn at Cincinnati and on the roads using these tires, they have done well,—as well as any other tires made in this country. These tires are hammered throughout, and bent, welded and blocked to a good finish.

The finishing shops are well supplied with tools,—some of English construction and some built in Cincinnati. The large planing machine built at the works, is, we believe, the largest in the West. It appears to be a good tool, strong and well finished. The large lathe was also built upon the spot and is a good piece of work.

The Little Miami road is largely supplied with engines from these works. The general construction of the passenger engines may be stated as follows: Outside connected, slightly inclined cylinders, solid frame, suspended or stationary link motion. Dimensions generally as follows: 14 inch cylinders; 20 inch stroke; 5 feet or $5\frac{1}{2}$ feet wheels. Steamports 14 inches by $1\frac{5}{16}$ inches. Boiler 40 inch shell, containing 122 copper tubes, 2 inches in diameter and 10 feet long. Fire-grate 4 feet long and 38 inches wide. Running generally with $2\frac{3}{4}$ inch exhaust nozzles. On most of the passenger engines on the Little Miami road, the lead on steam port is $\frac{1}{8}$ inch (constant on all the notches) lead on exhaust $\frac{1}{2}$ inch;—the outside lap being $\frac{3}{4}$ inches and the inside lap $\frac{3}{8}$ inch.

These engines are doing remarkably good service on the Little Miami road. The ruling grades of that road are said to be 45 feet, but whether this is over-estimated or not, these engines are drawing trains daily of *nine* eight wheel cars, passenger and baggage. Trains of this size are drawn from Cincinnati to Loveland's, the latter being the commencement of the Hillsboro' road. The train leaves three cars here and goes on with six cars. It is to be remembered that these engines weight but about 42,000 pounds, and as they make abundance of steam with $2\frac{3}{8}$ inch nozzles and generally make 25 miles per hour over the grades, the performance must be admitted to be good. One of the same class of engines, with 5 feet wheel has taken 32 eight wheel loaded freight cars over what is rated as a forty feet grade. The engine was said to have carried but 120 pounds of steam, but that is impossible or else the grade is overrated. Still, it was a great feat, with so light an engine, to get such a load over the grade at all, the grade being a long one.

A freight engine lately built by Moore & Richardson, for the Little Miami road, and called the "Wm. McCammon," has performed quite well. It is an outside connected, ten wheel engine, 15 inch (fifteen) cylinder, 22 inch stroke and six 4½ feet drivers, the extreme forward and back pairs being 12½ feet apart centers. The whole weight of engine is 53,000 pounds, 40,000 pounds being on the drivers. The boiler is 44 inch shell, and has 122 two inch tubes, 10 feet 6 inches long; fire grate 52 inches long, and 38 inches wide. The valve motion is that of the shifting link. The tires are chilled cast iron. This engine is very strongly put together. As for its performance, we only regret that some of its trips were not reported to us in a shape warranting us in publishing them. That it does remarkably good service there is no doubt, although it is impossible, in the nature of things, that it should do all that we have heard reported of it. That we may not be charged with prejudice, we will say what we mean by this remark.

It is said that this engine with 100 pounds of steam has drawn 43 loaded 8 wheel cars, averaging 30,000 pounds each, over a 45 *feet grade*. The whole tractive power of an engine of this size, *with* 100 *pounds of steam*, cannot be but 9170 pounds. The gravity alone of 685 tons, (30 cars at 15 tons and engine and tender at 40 tons) on a 45 feet grade, is 11,645 pounds, or 25 per cent. more than the power of the engine. Adding the friction of the train, which could not possibly have been less than 5 pounds per ton, and the whole power required for the load named would be nearly two-thirds more than the engine could have with a pressure of 100 pounds.

The fact that this engine has been unthinkingly complimented beyond its real merits does not diminish its substantial claims to being reckoned as a good machine. While we know a disinterested and very intelligent railroad man who has pronounced it to be the model freight engine of Ohio, we, at the same time, wish to save it from injudicious and extravagant assertions of a degree of superiority which cannot exist in it. [From the *Railroad Advocate* (January 12, 1856), vol. 2, p. 1.]

MOORE & RICHARDSON'S ENGINES

These builders are keeping pace with the rest in improvements, and like each of the other builders, in an original way. We are glad to say a good word for compilers of machinery. Compiling shows a more independent and progressive spirit than copying outright, and at the same time it exhibits a respect and appreciation as to the labors and opinions of others, a feeling which is sure to come, sooner or later. The absence of this reliance on other men's experience has been the fruitful cause of abortions and failures. Each one of our readers knows some instance to prove this remark.

The acknowledged *points* of Moore & Richardson's machines are good boilers and strong work. Elegance and perfection of detail are not made a *cardinal* point as with some Eastern builders, and yet there is nothing ugly or ungainly about the machines.

For a 15 x 22 inch cylinder, the firebox is 52 x 41 inches. The barrel is 42 inches diam. at the front end, and 43 at the back end. There are 122 two inch flues, 10 feet long and ¾ inch apart. The good distance between the flues is an important item, not half well enough understood by builders generally.

The frame consists of a bar 4 x 2¼ inches, running straight from breast to back beam with pedestals forged on. Another bar, or strap, from just behind the cylinders, extends to the foot of the front pedestal and another connects the pedestals, besides the jaw braces, which are secured by lugs and bolts. The lower strap is joined to the upper one behind the cylinder, by a key let into both, and by 3 bolts. A heavy angle iron connecting the frames in front, holds the wooden breast beam. The center and outside bearing truck, a very strong though somewhat complicated device, will be described at another time.

The equalizing beam is under the main frame, and the springs above, as usual. The Rogers expansion brace is employed on the fire-box. The center line of truck is 4 inches in front of the forward end of the smoke arch, thus throwing more weight than usual on the drivers. The foot board is of boiler plate, and the back braces running to it from the fire-box, are unusually heavy. No provision is made for expansion, as is sometimes the case, hence the boiler rises as it expands. It is questionable whether there is any virtue in a back brace which *has* an expansion apparatus. The slide rod yoke is of cast iron, with a leg to rest on each bar of the frame, to which it is securely bolted.

The cylinders are not level, being over the back truck wheels, instead of between the wheels. The cross head is hollow, and the rods flat, and much smaller at the ends (inside the heads) than in the middle. Some builders make them straight. Here is another open question. They are forged in a *solid bar*, no welding or scrap work being allowed.

The cylinder fastening is light and strong. The smoke arch being round, a flange from about the center of the cylinder projects straight towards it, and another from the valve seat, is parallel with the first, just above it. The two flanges are joined by the steam and exhaust pipes, which are cast between them, and by a heavy flange or plate, embracing a portion of the smoke arch. The frame is immediately under the lower flange, and holds it fast by lugs and wedges, as well as bolts.

The valve gear is the suspended link. The eccentric rods are 46 inches long, and the radius of the link 42 inches, which of course is the length of the valve connection from the center of link to center of rock shaft. The link is 10¾ inches center to center of knuckle joints, the hangers 15¼ inches, and the lifting arms 7¼ inches long. The center of tumbling shaft is 12½ inches above the center of engine, and directly over the center of link, which is suspended in the center both ways. The lap is ¾, and the eccentric lead ⅞ inch, hence the valve lead at full throw is ⅛ inch. The exhaust when the engine is on the point of taking steam, has ⅜ inch lead or clearance. The valve has 4¾ inch throw at full stroke, and opens ⅞ inch when cutting off at half stroke. The steam ports are 13 x ¹⁵⁄₁₆ inch, and the exhaust 13 x 2¾ inches.

The dome is in the middle of the boiler, and the dry pipe is secured to the steam pipe by a strap passing through lugs on its sides, and under the steam pipe. This pipe has a cone joint in the flue sheet. The branch pipes are secured to the T piece by flanges and bolts as of old.

The pump is of the modern sort, viz. a flat faced cup valve in a cage. The check valve is in the center of the front sheet, and consists of a valve box, elbow pipe and flange cast together. The valve seat and another elbow holding a packing box for the copper pipe, are secured to the first casting, by 3 bolts.

The cab is rather showy, and quite substantial. The running board extends only over the top of the wheel guards.

Moore & Richardson make their own tire, and find it to be a reliable article, though it lacks the finish of the great American tire. They also make all their forgings, employing for that purpose a Nasmyth hammer. They do *not* use nor approve of the use of *scrap*, for railroad forgings. Experience has taught them that it cannot be relied upon, however carefully made. They are very particular in their forging, and avoid welds as much as possible. The shapes for pedestals are made from a single bar

A very extensive steamboat business is carried on at this establishment, which is one of the oldest in the West. The boilers, as before observed, are of excellent quality and shape. On the whole, the work will compare favorably in point of strength with general locomotive work of the first class. [From the *Railroad Advocate* (September 6, 1856), vol. 3, p. 3.]

Appendix 2

The following chapter from *Trips in the Life of a Locomotive Engineer* (pages 131–136) pictures a journey in the cab of a Cincinnati-built locomotive about 1860. The engine described is the *D. W. Deshler*, built by Moore & Richardson in 1854 for the Little Miami Railroad.

FORTY-TWO MILES PER HOUR

Nearly every person that we hear speak of travel by rail, thinks that he has, on numerous occasions, traveled at the rate of sixty miles an hour; but among engineers this is known to be an extremely rare occurrence. I myself have run some pretty fast machines, and never had much fear as to "letting them out," and I never attained that speed for more than a mile or two on a down grade, and with a light train, excepting on one or two occasions. Supposing, however, reader, that we look a little into what an engine has to do in order to run a mile in a minute, or more time. Say we go down to the depot, and take a ride on this Morning Express, which goes to Columbus in one hour and thirty-five minutes, making two stops. We will get aboard of the Deshler, one of the smartest engines on the road, originally built by Moore & Richardson, but since then thoroughly over-hauled, and in fact rejuvenated, by that prince of *master*-mechanics, "Dick Bromley." And you may be sure she is in good trim for good work, as it is a habit with Dick to have his engines all so. She is run by that little fellow you see there, always looking good-natured, but getting around his engine pretty fast. That is "Johnny Andrews," and you can warrant that if Dick Bromley builds an engine, and Johnny runs her, and you ride behind her, you will have a pretty fast ride if the time demands it. The train is seven minutes behind time to-day, reducing the time to Columbus—55 miles— to one hour and twenty-eight minutes, and that with this heavy train of ten cars, all fully loaded. After deducting nine minutes more, that will un-doubtedly be lost in making two stops, this will demand a speed of forty-two miles per hour; which I rather guess will satisfy you. You see the tender is piled full of wood, enough to last your kitchen fire for quite a while; but that has got to be filled again; for, ere we reach Columbus, we shall need two cords and a half. Look into the tank; you see it is full of water; but we shall have to take some more; for between here and Columbus, 1558 gallons of water must be flashed into steam, and sent traveling through the cylinders.

But we are off; you see this hill is before us; and looking behind, you will see that another engine is helping us. Notwithstanding that help, let us see what the Deshler is doing, and how Johnny manages her. She is carrying a head of steam which exerts on every square inch of the internal surface of the boiler, a pressure of 120 pounds. Take a glance at the size of the boiler; it is 17 feet 6 inches long, and 40 inches in diameter. Inside of it there is the fire-box, 48 inches long, 62 inches deep, and 36 across. From this to the front of the engine, you see a lot of flues running. There are 112 of these, 10 feet 6 inches long, and two inches in diameter; and of the inner surface of all this, every square inch is subjected to the aforesaid pressure, which amounts to a pressure of 95,005 pounds on each flue. Don't you think, if there is a weak place anywhere in this boiler, it will be mighty apt to give out? And if it does, and this enormous power is let loose at once, where will you and I go to? Don't be afraid, though; for *this* boiler is built strongly; every plate is right and sound. Open that fire-door. Do you hear that enormously loud cough? That is the noise made by the escape, through an opening of 31 square inches only, of the steam which has been at work in the cylinder. You can feel how it shakes the whole engine. And see how it stirs up the fire. Whew! isn't that rather a hot-looking hole? The heat there is about 2800° Centigrade scale.[*] But we begin to go faster. Listen! try if you can count the sounds made by the escaping steam, which we call the "exhaust." No, you cannot; but at every one of those sounds, two solid feet of steam has been taken from the boilers, used in the cylinder, where it exerted on the piston, which is fourteen inches in diameter, a pressure of nine tons, and then let out into the air, making, in so doing, that noise. There are four of those "exhausts" to every revolution of the driving-wheels, during which revolution we advance only 17⅔ feet. Now we are up to our speed, making 208 revolutions, changing 33⅓ gallons of water into steam every minute we run, and burning eight solid feet of wood.

We are now running a mile in one minute and twenty-six seconds; the driving-wheels are revolving a little more than 3½ times in each second; and steam is admitted into, and escapes from, the cylinders fifteen times in a second, exerting each time a force of nearly nine tons on the pistons. We advance 61 feet per second. Our engine weighs 22 tons; our tender about 17 tons; and each car in the train with passengers, about 17 tons;

*This temperature was apparently in error for iron melts at 1535° C.

so that our whole train weighs, at a rough calculation, 209 tons, and should we strike an object sufficiently heavy to resist us, we would exert upon it a momentum of 12,749 tons—a force hard to resist!

Look out at the driving-wheels; see how swiftly they revolve. Those parallel rods, that connect the drivers, each weighing nearly 150 pounds, are slung around at the rate of 210 times a minute. Don't you think that enough is required of an engine to run 42 miles per hour, without making it gain 18 miles in that time? Those tender-wheels, too, have been turning pretty lively meanwhile—no less than 600 times per minute. Each piston has, in each minute we have traveled, moved about 700 feet. So you see that, all around, we have traveled pretty fast, and here we are in Columbus, "on time;" and I take it you are satisfied with 42 miles per hour, and will never hereafter ask for 60.

Let us sum up, and then bid good-bye to the Deshler and her accommodating runner, Johnny Andrews. The drivers have revolved 16,830 times. Steam has entered and been ejected from the cylinders 67,320 times. Each piston has traveled 47,766 feet, and we have run only 55 miles, at the rate of 42 miles per hour.

Appendix 3

J. L. Whetstone (see pages 100-102) published a precise mechanical description of his radial valve gear in the *Journal of the Franklin Institute* in February 1882. The following extract from this article discusses his unconventional valve motion as used on the Sellers Panama locomotives and some of the engines built by Niles & Company

From the peculiarity of the general arrangement of the machinery it was found impracticable to use more than one eccentric for operating the valves of each engine, and it was necessary to use a valve with considerable lap. There being barely room for one eccentric for each engine on the driving axle, the device of shifting a lead eccentric across the axle for the purpose of obtaining lead for the forward and backward movements could not be applied. The valve gear which was finally adopted was substantially the same as represented in Fig. 7,[*] with slight modifications. The eccentric was set so as to be at half throw when the crank pin was at the ends of the stroke or at the dead point, and connected by a rod to an arm on the rocker shaft having at its other end a double arm, carrying a link bar for the purpose of giving reversing movements to the valve. The position of the rocker arm and link bar was therefore the same when the crank was at either end of the stroke, viz., that shown in the Fig. 7, at A B C. For the purpose of giving the advanced movement requisite for the lap and lead of the valve, the shifting or reversing rod (one end of which is properly swiveled to the link bar) is connected to the fulcrum F of a lever, the longer arm of which is suitably attached to the cross-head of the engine, the shorter arm being geared to the valve rod at G, or to any suitable device necessary to transmit the movement to the valve. The length of the arms of this lead-lever are such that when the fulcrum F is at the half throw of the eccentric the upper wrist of the lever G is removed from the centre line II K to the extent of the lap and lead of the

---

*Reproduced as figure 42 on p. 101.

135

valve, the longer end connected with the crosshead then at one end of the stroke, and if the crosshead be at the other end of the stroke, the upper wrist will be as far removed to the opposite side of the centre line.

It will also be observed that the position of the wrist operating the valve rod will remain the same, in whatever part of the link bar the reversing bar of rod D F may be situated, whether in full gear forward or backward, or at any intermediate point, and the lead of the valve will be the same at both ends of the stroke. The throw of the eccentric for this valve gear will be shorter than the travel of the valve, inasmuch as part of the valve movement is obtained from the crosshead. In practice it is found that about two-thirds of the lead of the valve is obtained from the crosshead, and the eccentricity of the eccentric is lessened to that extent. At first sight it would seem as though the whole was due to the cross-head, but it must be borne in mind that during the last half of each stroke of the piston the eccentric motion is in a direction opposite to that of the crosshead, thus combining the two movements to extend the travel of the valve. The effect of this combination is to accelerate the movement of the valve at the opening of the ports, and to retard it at and towards the end of the throw of the valve, thus giving a longer admission of steam with a given lap of valve than by the eccentric motion alone. It would be quite practicable to operate the rocker arm link of one engine from the crosshead of the opposite engine, the lead being obtained from its own crosshead, and probably the greatest objection to such an arrangement would arise from the fact that the disability of one engine through an accident would render the valve movement of the other engine inoperative.

136

Appendix 4

The *Railroad Advocate* in 1856 published these two accounts of Niles & Company locomotives.

WESTERN LOCOMOTIVES

Niles & Co., as indeed all the other builders in Ohio, have enough intelligent liberality to throw open to the public the principal dimensions and details of their engines. The time was past, long ago, when locomotive builders could arbitrarily impose their prejudices upon their customers, and make secrets out of their proportions and arrangements of machinery. Purchasers, and not builders of engines, now dictate the general direction of improvement in locomotives, and insist upon sizes and constructions which are founded in a common experience, or in broad or closely contested opinion. It is just, therefore, more to the railroad public than to the builders, that the details of the engines which now form the standard manufacture of the West should be given at large. The present article will afford no more room than for a description of Niles & Co's. engines, but memoranda of those by some of the other Ohio builders will be given elsewhere in today's paper.

An engine built by Niles & Co. for the Indianapolis and Cincinnati railroad may be taken as the subject of illustration. Outside connected, nearly level cylinders, spread truck, link motion, wagon top boiler, solid frame. The boiler is 44 inches diameter in the shell and contains 141 copper tubes, 2 inches outside diameter and 10 feet 10 inches long. The fire-box is 50 inches long on the grate, 38 inches wide, and 58 inches deep. There is a good sized dome over the firebox, which is considerably elevated in the crown to give ample steam room. The furnace is stayed very strong.

The cylinder is 15 inches diameter with 22 inches stroke of piston. Four drivers 5½ feet diameter. The steam ports are the longest which we have seen on any locomotive,—viz., 18½ inches, width 1 inch. The greatest throw of valve is 4¾ inches; outside lap ¾ inch, and inside lap ³⁄₁₆ inch, on each end. The cylinders are thoroughly jacketed with brass. The cylinder fastening is made by bolting directly (without a saddle) upon a round smoke box. The flat flange which bolts to the frame extends inward to just under the middle of the smoke-box, and there turns up a vertical flange reaching to the smoke-box. The valve seat flange also reaches horizontally to the side of the smoke-box, and an apron, or curved flange, extends around the smoke-box between these flanges. The two cylinders are bolted together through the upright flanges under the middle of the smoke-box, while with the fastening to the smoke-box and to the frame, the job is made very complete.

The frames are solid with pedestals or jaws welded on. The jaws have tightening wedges for taking up the wear on the sides of the boxes. Forward of the front jaw, the frame makes an offset downwards, as usual. From the front leg of the front jaw, at about half its height, a brace is bolted on and reaches to the under side of the offset, where it fastens on. From the bottom toe or lug of the same jaw, a long brace extends to the front end of the frame, trussed, at intervals, to the frame. The frame being of good size, and stiffly braced to the boiler, the whole structure is very stiff. The fire-box is fastened to the frame by an angle iron, as is usual, the bolt holes through the angle iron being slotted or oval, to allow for the expansion of the boiler when hot. A long shallow groove or key-way is planed in the top of the frame, under the angle iron, and a corresponding groove is planed out of the bottom of the angle iron. A long key is then driven in, to take whatever side strain there may be without crowding on the bolts.

The springs are long and easy, and between the lower ends of the outer spring straps and the frame, blocks of rubber are interposed.

The drivers are 7 feet between centers,—a good spread and ensuring a liberal adhesive weight. The drivers are 5½ feet diameter.

The link motion is suspended upon a new principle, discovered by Mr. John L. Whetstone, Engineer of the Works. We are not at liberty to disclose the manner in which the centers of suspension are found by Mr. Whetstone's plan, but it is very simple and solves the great problem, so long pursued, of a perfectly equal admission, cut-off and exhaust, and equal lead, on both strokes and in every notch of forward and back gear. We say this understandingly, having had the whole process of finding these centers, demonstrated to us, by diagrams and working models. The principle is beautiful, and the locomotive builders of this country could well afford to unite in a handsome reward to Mr. Whetstone, for the simple process by which he secures the result to which so many unsuccessful efforts have been directed.

The truck frames are center bearing and the wheels are 4 feet 6 inches apart centers. The journals are inside the wheels (that is McQueen's double bearings are not adopted). The frame is very light, very simple, and self-sustaining throughout. It is not a "square truck frame," so called, but there are two side frames or bars, 4 by 1⅜ inches, and having each two pedestals welded on solid, as in Brandt's engines. The peculiar feature of the truck is in the way in which these sides are cross braced to each other. A tapering cross-beam, 7 by 1¼ inches at the center, is placed transversely across the frame, midway between the jaws, and is bolted to the top of the frame. The cast iron center-pin bushing rests on top of this beam and flanges coming down from the bottom of the bushing embrace the sides of the beam. On the outside of these flanges, braces of flat iron are bolted on, edge upwards, and are bent out to reach to the backs of the several jaws. There are transverse braces extending across to stiffen the jaws together. The springs and equalizing levers, and the truss bolts and thimbles between the jaws are the same as in Eastern engines. Glancing at random over the engine, the boiler is very low down; the pumps are of good capacity and with air chambers on the suction and forcing sides; the rockers are of wrought iron, with two bearings to each; the links are stout and well got up; the outside trimmings are in good taste; the bolting is generally strong, the bolts being of good size and turned and drove in reamed holes; the bearings are protected against wear, generally by allowing large surfaces and by thorough case-hardening at all points where that process is admissable.

These engines were built after the designs and under the superintendence of John L. Whetstone, Esq., assisted by Mr. Coleman Sellers and Mr. O. H. P. Little. The workmanship throughout is highly creditable to the skill and taste of the mechanics, as also to the the enterprise and liberality of the proprietors of the works.

These engines, we are told, have given a good account of themselves, for speed and power, side by side, with other crack engines from some of the first class shops of the country. Without detracting from anybody's claims, the Niles engines are creditable to Western skill, and the talent and resources of the works of these proprietors are such as Western men may fairly be proud of. [From the *Railroad Advocate* (January 12, 1856), vol. 2, p. 3.]

NILES & CO.'S NEW ENGINES

The appearance, if not the quality of these machines, has improved of late. The elegant lithograph recently issued by Niles & Co. does not misrepresent their works. The engines embody various decided improvements, fully proving that one head cannot do all the invention. Mr. WHETSTONE, the designer, draftsman, and practically the foreman, is a very thorough and valuable mechanic and thinker. It is a very en-

139

couraging thought to the railroad community, that so many able competitors are found in all our shops. Our readers can judge of relative merit, from descriptions of the novelties of each builder. Niles & Co's recent 16 inch cylinder engines are as follows:

The framing is something after the McQueen fashion, and consists of 2 bars, the upper one 4 x 2¼ inches and the lower one 4 x ⅞ inches, running from the breast beam, parallel to within some 3 feet of the forward driver center, where they separate, passing respectively above and below the pedestals which are forged solid with the upper bar. This bar above the pedestals, is 3¾ x 3⅛ inches. Forward, the 2 bars of the frame are joined by bolts passing through thimbles in the shape of braces. The strap under the jaws has lugs forged on, which project up in front and behind each pedestal. The front brace rests against a lug turned up on the front end of the top bar. A back jaw brace runs to the back end of the frame. The expansion apparatus is novel and simple. It will be described at another time. The foot board is cast iron, very solid. A strong foot board is very important to hold the frames. Such a simple proposition, however, has its opponents. The equalizing beam is under the main frame, and the centers of springs ⁵⁄₁₆ inch outside the vertical center of the jaws, thus allowing the fire-box to be unusually wide, and leaving a 3¾ inch space between it and the tire of the drivers. This is not a bad arrangement. The eccentric suspension of the springs being so slight, does not cause any unsteadiness. Rubber blocks are placed as usual under the hangers.

The boiler is of the ordinary raised crown sheet style, as introduced by Rogers. The grate is 40 x 50 inches. The barrel is 45 inches in diam., and contains 152 two inch flues 11 feet long. The dome is over the firebox.

The cylinder fastening is peculiar and not very heavy. It is difficult to pronounce on its comparative merits without figuring up the weight and strain, to do which we have not at present the data. A heavy flange, as long as the cylinder, projecting out horizontally from the center line of the cylinder to the center line of the smoke arch meets the corresponding flange of the other cylinder. The two then run up vertically to the bottom of the smoke arch and are bolted together. Another flange from the top of the valve seat projects parallel to the first to the center of the bottom of the smoke arch, being scooped out a little to receive the arch. This flange then continues up a few inches on the side of the arch to receive bolts. Between the two horizontal flanges are the steam and exhaust pipes, the whole arrangement, of course, being in one casting. Under the lower horizontal flange, close to the cylinder, is the frame. A ball joint and single bolt hold the steam pipes to the cylinder. This is like Mason's, except that the bolt is diagonal and projects through the side of the pipe, whereas in Mason's the bolt is horizontal and projects through the elbow of the pipe, the nut being set up outside the smoke arch. The T piece has one bolt and a ball joint much like Mason's.

The valve motion is the suspended link, and at least the suspension of the link is entirely original. This is the only valve which cuts off square all round. We are not at liberty to tell how the centers are laid out, neither are we persuaded that a perfectly equable cut off and a considrable or even a moderate slip is preferable to a motion opening the exhaust square at all times, varying the cut off slightly, and allowing no slip of the block at points where the valve is usually cutting off, particularly when such a cut off necessitates an increased amount of machinery. But a perfect cut off is a highly praiseworthy, scientific, and workmanlike achievement, after all.

The tumbling shaft is suspended from $14\frac{1}{2}$ to $15\frac{1}{4}$ inches back of the vertical center line of the rocker arm, and from $2\frac{7}{8}$ to 4 inches below the rocker arm pin. Were the tumbling shaft hangers dropped from the frame as usual, the shaft would pass between the forward and back eccentric rod, hence hangers to suspend the shaft between the two sets of eccentric rods are necessary. To accomplish this, a brace is dropped from the frame, and an offset at its lower end to pass under the eccentric rods, projects up inside of the rods and holds the end of the shaft. All this work is securely braced.

The eccentrics have $4\frac{3}{4}$ and 5 inch throw, and the valve $\frac{3}{4}$ lap, and $\frac{1}{16}$ lead. The radius of the link is 6 feet, i.e., the distance from center of shaft, to center of rock-shaft. It is of the skeleton variety, being 2 inches thick, and $1\frac{1}{2}$ inches broad, in the narrowest places. It is suspended vertically, in the center, and horizontally, from $\frac{5}{8}$ to $\frac{7}{8}$ inch back of the center, thus giving of course more slip on the block than with the Rogers, B. K. & Co. etc. links, which are suspended $\frac{3}{4}$ inch back of center, and much more than Mason's which do not slip at all, at the ordinary running point. We mention this as a fact not an objection, for we presume the jury would not agree on the subject. The suspenders are 11 inches long. Their greatest width is across the engine, not lengthwise with it, as in almost, if not all other cases. This is a very sensible idea, and ought to have been adopted universally long ago. There is no strain which requires strength of the suspender in the direction in which the link moves. A half inch rod would hold it up. The link being suspended on one side, tends to careen sidewise, and hence the suspender should be strongest in the direction lengthwise with its pins. This is particularly economical too, because there must be a lug forged on the side of both ends of the suspender to make long bearings for the pins, and in this case no lugs are necessary, for the iron is already wide enough to form the bearing. In these machines it is 3 inches wide. The lifting arm is 17 inches long,

141

and the distance from center to center of knuckle joints is 12 inches. The rod has a solid head (no other joint) and a case hardened box, with a single key and set screw. The valve is moved by a yoke, in which the rod is fastened by nuts. The yoke does not in this case prevent the evil it is usually designed to avert,—the stripping of valve stems.

The length of the ports is the great charm of Niles & Co's. engines. The port for 15 and 16 inch cylinders is 1 x 18 inches. If 18 square inches of port area astonishes the fearful, they will at least acknowledge that length, and not width, is the important dimensions of a given port. If an opening contains 12 square inches, and is 8 inches long and 1¼ inches wide, when the valve cuts off at say 9 inches, and opens half an inch, the total opening will be 4 sq. inches. But if the same area is spread out into 12 inches long by 1 inch wide, a valve opening of ½ inch will give a total opening of 6 sq. inches, when cutting off at 9 inches. Now if an engine running at 40 miles an hour so wire draws her steam that a cylinder pressure of 60 lbs. is attained out of a boiler pressure of 100, with a 6 inch opening, what will the cylinder pressure be with a 4 inch opening? Diagrams tell a story about narrow ports, that has not been generally understood.

Again, with a short port, a greater lap and lead are required to produce the same effect on the piston. The ports we have described with a ¾ inch lap, will do as well as the ordinary port with an inch lap. But what is the objection to inch lap? A long valve, a large cup in the valve, a great pressure over the cup to consume power and wear out machinery, is one objection. There are others which we shall consider at another time.

Two words more, however, will tell us how we can have the boiler pressure in the cylinder, and light machinery, and preserved power;— BALANCED VALVES.

The slides of the Niles machines are held up after one of the Norris styles. A thimble is placed between the back ends of each of the rods, and a brace from the side of the barrel, forming a ∩ at the bottom rests on their top, a bolt passing through the thimble and ∩ on each pair. Each rock-shaft has two boxes, one on the slide and one on the frame, the valve arm being between the two. Suppose the slides to get out of line by wear or accident, is the box on the slide to be refitted? There is longer leverage, however, and less twist and wear in the boxes, with this arrangement.

The pump has the ordinary flat valve, is not ugly to look at, and is said to work well. Speaking of pumps, any decent pump will work well enough, if the valves do not raise too high, or leak. The check valve is simple and well arranged. Some builders make a very expensive edifice on the side of the barrel, just to take in water. This is a mere elbow pipe, with a valve seat secured at the bottom with two bolts, and a thread cut on it to receive the hose coupling.

142

The driving boxes are novel, and good, being after the plan of Z. H. Mann, of Cincinnati.—On the front of the brass box is a separate piece, which is set up, as it wears, by a wedge passing between it and the inside of the cast iron box. The wedge is adjusted by a rod, and nuts on either side of the jaw brace. The iron boxes are planed up so as to place the center of the shaft in the exact center of the jaw. This is done by slipping a ring of the size of the shaft into the box, while it is on the planer, dropping the chisel to touch the ring, and then planing the slot where the box embraces the jaw, to just this depth. The same operation with the other slot, accomplishes the object.

There are various miscellaneous parts worthy of note. The cab is rather showy, and very commodious. The ornamental brass work is well done but not elaborate, and not, in all cases, graceful. The running board is very light, and of the modern style,—not resembling a railroad bridge as much as some running boards we have seen. The cow-catcher is equally showy and awkward. The painting is very neat, and not gaudy.

The truck is novel, and very good. We shall describe it at another time.

On the whole, NILES & Co. built a first-rate engine; and it will compare, in work at least, with the general build of the East. [From the *Railroad Advocate* (September 20, 1856), vol. 3, p. 2.]

Appendix 5

In 1856 the *Railroad Advocate* published the following article comparing the performances of passenger engines operated by the Cincinnati, Hamilton, and Dayton Railroad. The line's Niles-built locomotive, the *S. Gebhart*, made an impressive showing.

The equipment of the Cincinnati, Hamilton, and Dayton road comprises engines from the best builders of the country. Rogers, the Taunton Co., and Hinkley, besides two Cincinnati firms, have severally turned out their best samples of work, upon the road under notice. The Taunton Co. have as good engines on the road as they ever built, and refer to them as such when negotiating with new roads in that neighborhood. Hinkley has but one engine on the road, the "L'Hommedieu,"—but it was originally named for himself—the "Hinkley,"—and was turned out as his best specimen of design and workmanship. It is the engine which has made the 60 miles of the Hamilton and Dayton road in something less than 75 minutes,— taking a good train over a 26 feet grade.

We allude to these particulars as facts merely, and in order to give just credit to a Cincinnati-built engine,—the "S. Gebhart,"—the only one on the road from Niles & Co's. shops.

The following comparison will show the service and expenses of all the principal passenger engines in use on the road, during the year 1855. [See table on p. 146.]

It will be seen that the "S. Gebhart"—the Niles' engine—came within but 35 miles of reaching the highest mileage of any engine on the road,— that its repairs hardly exceeded the average cost for that of all the others,— and that it was thirty-six per cent. more economical of fuel than any other engine run,—and from fifty to one hundred and fifty per cent. more economical of fuel than several of the other engines named!

When we visited Cincinnati, last winter, we were struck with the liberal design, excellent workmanship, and especially, the exact mode of fitting the work together in the "setting-up-shop." We had not expected to see so decided an improvement on the work turned out in Cincinnati only two or three years before.

Names of Locomotives	Names of Builders	Cost of Repairs	No. of Miles Run	Cords of Wood	Miles per Cord of Wood
S. Gebhart	Niles & Co.	$1,549	31,695	595	53
A. M. Taylor	Taunton Works	1,759	31,730	810	39
Glendale	Rogers	1,208	27,865	711	39
E. B. Reeder	Taunton Works	1,815	26,470	707	37
L'Hommedieu	Boston Works	847	22,850	627	36
Cincinnati	Harkness & More [sic]	916	22,255	690	32
Carrolton	Rogers	2,287	17,440	474	37
Carlisle	Rogers	1,911	13,765	400	34
Franklin	Taunton Works	1,341	12,275	390	31
J. Woods	Harkness & Co.	1,574	10,020	492	22

The Cincinnati builders deserve great credit for their liberal enterprise and indomitable perseverance in working against all the difficulties incident to a new business,—starting at the outset with inexperienced hands, presumptuous designers, and uncertain materials,—but conquering these circumstances, and rising in merit and reputation to deserve a patronage limited only by the great demand for engines in the West.

There are many excellent points in Messrs. Niles' engines—the heating surface and boiler content, the very liberal width of steam passages (18½ inch ports for a 15 inch cylinder) and the exact equality of admission, lead, and expansion, attained by their peculiar suspension of the shifting-link. The engines are original in design, but are not based on ultra ideas of proportion or arrangement; while they also combine nearly every known means for the perfect production and distribution of steam. [From the *Railroad Advocate* (June 7, 1856), vol. 3, p. 1.]

146

Appendix 6

List of Locomotives
Built in Cincinnati 1845-1868

The following list of locomotives built in Cincinnati has been compiled from annual reports of the railroads concerned. It is incomplete but certainly accounts for a large percentage of the engines constructed by the Cincinnati builders. Unfortunately, most railroads stopped listing their equipment after about 1860, so that the information presented for the following years is scanty.

ANTHONY HARKNESS & SON 1845-1849

Railroad	Name or number	Date delivered	Weight (tons)	Cylinders (inches)	Wheel diameter (inches)	Remarks
Little Miami	Bull of the Woods	1845 or 1846				No proof of existence. Not on official roster
"	Cincinnati	Nov. 1846	16			
"	Hamilton	March 1847	16		54	4–4–0. "Entirely rebuilt 1849"
"	Warren	May 1847	14			Scrapped 1857
"	Greene	June 1847	15		60	4–4–0
"	Buckeye	June 1848	18½			
"	Gen. Harrison	June 1848	15			Sold to Nashville & Chattanooga RR. May 1851

147

Railroad	Name or number	Date delivered	Weight (tons)	Cylinders (inches)	Wheel diameter (inches)	Remarks
Little Miami	*Simon Kenton*	June 1848	15			
"	*Shawnee*	Jan. 1849	15			
Madison & Indianapolis	*Johnson (9)*	1847				Out of service by 1860
"	*Jefferson (10)*	1847				Out of service by 1860
"	*Jennings (12)*	1848				Out of service by 1860
"	*J.F.D. Lanter (13)*	1848				
"	*Gov. Witcomb (14)*	1848				
Mad River & Lake Erie	*Urbana*	1847	12	12 x 18	54	4-4-0
"	*Bellefontaine*	1847	12	12 x 18		
"	*Springfield*	1849	16	12 x 18	60	4-4-0
"	*Belle Center*	184?		12 x 18	62	4-4-0. Purchased by Cleveland, Columbus & Cincinnati RR. April 1852

ANTHONY HARKNESS & SON 1850

Railroad	Name or number	Date delivered	Weight (tons)	Cylinders (inches)	Wheel diameter (inches)	Remarks
Columbus & Xenia	*Washington*	Feb.	19¼		72 (?)	Sold to Bellefontaine & Indiana RR. 1851
Little Miami	*Gov. Morrow*	Jan.	19			First *Gov. Morrow* built by Rogers 1841 was traded for this new locomotive
"	*Columbus*	May	19		54	
"	*Hercules*	Dec.	23		48	4-6-0
Nashville & Chattanooga	*Tennessee*	Dec.	20			4-4-0
"	(?)					No data available

148

Railroad	Name or number	Date delivered	Weight (tons)	Cylinders (inches)	Wheel diameter (inches)	Remarks
ANTHONY HARKNESS & SON 1851						
Cincinnati, Hamilton & Dayton	J. B. Varnum	March	18½		54	4-4-0
"	Ethan Stone	June	18½		54	4-4-0. Scrapped 1858
"	Cincinnati	Sept.	19		66	4-4-0
"	Hamilton	Oct.	19		66	4-4-0
"	Eaton	Dec.	19		66	4-4-0
Little Miami	Fire Fly	Feb.				
"	Jupiter	March	23		48	4-6-0. Sold U.S. Military Railroads, 1862
"	Elk	April				
"	Xenia	Aug.	19		60	First Xenia built by Rogers 1844. Sold to Cleveland, Cincinnati & Columbus RR.
"	Atlas	Nov.	23		60	4-4-0
Nashville & Chattanooga	V. K. Stevenson	July				
ANTHONY HARKNESS & SON (HARKNESS, MOORE & CO.) 1852						
Central Ohio	Kenton	July	20			
Cincinnati, Hamilton & Dayton	John Johnson	March	24		60	4-4-0. Harkness, Moore & Co.
Little Miami	Antelope	March	20		60	
"	Gov. Morrow	July	24		60	Third Gov. Morrow?
"	Robert Fulton	Aug.	24		48	4-6-0
"	Simon Kenton	Nov.	24		48	4-6-0
Louisville & Frankfort	Tom Smith					4-6-0
"	John I. Jacobs					4-6-0

149

Railroad	Name or number	Date delivered	Weight (tons)	Cylinders (inches)	Wheel diameter (inches)	Remarks
Memphis & Charleston	Somerville	June	19		54	
"	J.W. Garth	June	16½		48	
"	Lurahoma	Dec.	23		48	
Nashville & Chattanooga	W.S. Watterson	Feb.				Outside frame
"	John Eakin	April				

MOORE & RICHARDSON 1853*

Railroad	Name or number	Date delivered	Weight (tons)	Cylinders (inches)	Wheel diameter (inches)	Remarks
Central Ohio	Reindeer	Jan.	22			
Cincinnati, Hamilton & Dayton	Trenton	Jan.	24		48	4-6-0. Harkness, Moore & Co.
"	Carthage	April	24		48	4-6-0. Harkness, Moore & Co.
"	S. Fosdick	June	23		60	4-4-0. Harkness, Moore & Co.
"	J.C. Wright	July	23		60	4-4-0
"	J. Woods	Aug.	23		60	4-4-0
"	Lockland	Sept.	24		48	4-6-0
Cincinnati, Wilmington & Zanesville	Lancaster	Nov.	21		58	Cost $8,272
"	New Lexington	Dec.	21		58	Cost $8,272
Covington & Lexington	Col. Morgan	March	22			4-4-0
Little Miami	Jno. Kilgour	May	19		60	
"	Robt. Neil	May	10		60	
"	Doctor Goodale	Oct.	19		60	4-4-0
"	W.S. Sullivant	Oct.	19		54	

---

*Moore and Richardson built the Ohio & Mississippi RR.'s Nos. 33 and 34; however, the date of construction and other data are unknown.

Railroad	Name or number	Date delivered	Weight (tons)	Cylinders (inches)	Wheel diameter (inches)	Remarks
Memphis & Charleston	*Magnolia*	March	23½		60	
"	*Antelope*	March	23½	14 x 20	60	
"	*Southerner*	Nov.	23½		60	
"	*Tuscumbia*	Nov.	22½	14 x 20	54	

MOORE & RICHARDSON 1854

Railroad	Name or number	Date delivered	Weight (tons)	Cylinders (inches)	Wheel diameter (inches)	Remarks
Central Ohio	*Elk*	July	22			
Little Miami	*D.W. Deshler*	Jan.	19½		66	4-4-0
"	*John Kugler*	Nov.	19		66	4-4-0
Little Miami & Columbus & Xenia	*Nat. Wright*	April	19½		66	4-4-0. Scrapped 1868
"	*Wm. Dennison, Jr.*	May	19½		66	4-4-0
"	*R.R. Springer*	Oct.	19½		66	4-4-0

MOORE & RICHARDSON 1855

Railroad	Name or number	Date delivered	Weight (tons)	Cylinders (inches)	Wheel diameter (inches)	Remarks
Covington & Lexington	*Fayette*	Aug.	18			
"	*Kenton*	Aug.	23			
Little Miami	*Wm. McCammon*	Sept.	23		54	4-6-0
"	*Jas. Hicks, Jr.*	Oct.	23		54	4-6-0
Memphis & Charleston	*Southerner*					Second *Southerner?*
Nashville & Chattanooga	*Daniel Webster*	March				
"	*Henry Clay*	March				
"	*John C. Calhoun*	March				
Indianapolis & Cincinnati	*Sam Wiggins*					4-4-0

151

Railroad	Name or number	Date delivered	Weight (tons)	Cylinders (inches)	Wheel diameter (inches)	Remarks
MOORE & RICHARDSON 1856						
Little Miami & Columbus & Xenia	Abraham Hirling	Dec.	23		54	4–6–0
Marietta & Cincinnati	Hillsboro	Jan.		14 x 20	60	4–4–0
Memphis & Charleston	Memphis	April	24¾		54	
"	J.F. Cooper	Nov.	20		60	
"	Pocahontas	Nov.	20		60	
"	Gray Eagle					
MOORE & RICHARDSON 1857						
Indianapolis & Cincinnati	H. C. Lord			15 x 22	62	4–4–0
"	Robert Meek			15 x 22	62	4–4–0
"	W. H. Clement			15 x 22	62	4–4–0
Little Miami & Columbus & Xenia	William B. Hubbard	Jan.	23		54	4–6–0
"	John Bacon	May	23		54	4–6–0
"	Jos. R. Swan	June	23		54	4–6–0
Louisville & Nashville	New Haven (5)	Sept.	22	15 x 20	60	4–4–0. Scrapped 1878
"	Marion (6)	Sept.	22	15 x 20	60	4–4–0
Memphis & Charleston	Chickasaw	Jan.	20		60	
"	Iuka	Jan.	24½		48	
"	Cherokee	May	24½		48	
"	Powhatan	May	24½		48	
MOORE & RICHARDSON 1858						
Louisville & Nashville	Hardin County	July	22	15 x 20	66	4–4–0. Scrapped 1876
"	Green River	Sept.	22	15 x 20	66	4–4–0. Scrapped 1880
"	James Guthrie		27⁹⁄₁₀	16 x 22	54	4–4–0. Sold U.S. Military Railroads 1863

Railroad	Name or number	Date delivered	Weight (tons)	Cylinders (inches)	Wheel diameter (inches)	Remarks
MOORE & RICHARDSON 1859						
Louisville & Nashville	*James F. Gamble*		27$\frac{9}{10}$	16 x 22	54	4–4–0. Sold U.S. Military Railroads 1863
"	*Edmonson*		27$\frac{9}{10}$	16 x 22	54	4–4–0. Scrapped 1885
"	*Barren*		27$\frac{9}{10}$	16 x 22	54	4–4–0. Sold U.S. Military Railroads 1863
"	*Warren*		27$\frac{1}{2}$	16 x 22	54	4–4–0. Scrapped 1885
"	*Simpson*		23$\frac{3}{5}$	15$\frac{1}{2}$ x 20	66	4–4–0. Sold 1881
"	*Quigley*		23$\frac{3}{5}$	15$\frac{1}{2}$ x 20	66	4–4–0. Scrapped 1885
"	*Newcomb*		23$\frac{3}{5}$	15$\frac{1}{2}$ x 20	66	4–4–0. Scrapped 1876
"	*22*	Oct.	23$\frac{3}{5}$	15$\frac{1}{2}$ x 20	66	4–4–0. Sold 1878
"	*23*		27$\frac{9}{10}$	16 x 22	60	4–4–0. Scrapped 1876
"	*24*		27$\frac{9}{10}$	16 x 22	60	4–4–0. Scrapped 1876
"	*25*		27$\frac{9}{10}$	16 x 22	60	4–4–0. Sold U.S. Military Railroads 1863
"	*26*		27$\frac{9}{10}$	16 x 22	60	4–4–0. Scrapped 1891
"	*27*		27$\frac{9}{10}$	16 x 22	60	4–4–0
Louisville, New Albany & Chicago	*Clipper*		19	13 x 20	62	Cost $7,500
MOORE & RICHARDSON 1860						
Louisville & Frankfort & Lexington & Frankfort	*Fayette*		26	15 x 22	54	4–4–0. Scrapped 1886. Possibly secondhand
Louisville & Nashville	*28*		27$\frac{9}{10}$	16 x 22	60	4–4–0. Scrapped 1876

Railroad	Name or number	Date delivered	Weight (tons)	Cylinders (inches)	Wheel diameter (inches)	Remarks
Louisville & Nashville	29		27⁹⁄10	16 x 22	60	4–4–0. Sold 1871
Louisville, New Albany & Chicago	Tornado		27	16 x 22	56½	

MOORE & RICHARDSON 1861

Railroad	Name or number	Date delivered	Weight (tons)	Cylinders (inches)	Wheel diameter (inches)	Remarks
Memphis & Charleston	Michigan					Railway & Locomotive Historical Society Bulletin 108, p. 75,
"	Tennessee					states both engines built "new" for U.S. Military RR.

MOORE & RICHARDSON 1864

Railroad	Name or number	Date delivered	Weight (tons)	Cylinders (inches)	Wheel diameter (inches)	Remarks
Chicago & Northwestern	Hercules					
Indianapolis & Cincinnati	25					Later Cleveland, Cincinnati, Chicago & St. Louis Railroad
"	29			16 x 22	62	4–4–0
Indianapolis, Cincinnati & Lafayette			32	16 x 22	62	Later Cleveland, Cincinnati, Chicago & St. Louis 238
Louisville & Nashville	65		30½	16 x 22	60	See Railway & Locomotive Historical Society Bulletin 5, p. 14
"	66		30½	16 x 22	62	See Railway & Locomotive Historical Society Bulletin 5, p. 14

154

Railroad	Name or number	Date delivered	Weight (tons)	Cylinders (inches)	Wheel diameter (inches)	Remarks
ROBERT MOORE & SON 1867						
Union Pacific	51			18 x 24	57	Scrapped 1898
"	54		43	18 x 24	57	4-6-0
"	55			18 x 24	57	4-6-0
"	56			18 x 24	57	To Omaha, Niobrara & Black Hills RR. 1880
NILES & COMPANY 1852*						
Cleveland & Pittsburgh	Stark	June		16 x 22	54	4-6-0
"	Columbiana	Aug.		16 x 22	54	4-6-0
Dayton & Western						No data. Possibly first Niles engine
Little Miami	Gray Eagle	Sept.	24		60	4-4-0
"	Gebhart	Sept.			54	4-4-0
"	Gov. Trimble	Nov.	19		60	4-4-0
"	Griffin Taylor	Dec.	19		54	4-4-0
Nashville & Chattanooga	Grampus	Nov.				
"	Col. Dickenson					
"	J. E. Thompson					
NILES & COMPANY 1853						
Bellefontaine & Indiana (Cleveland, Columbus, Cincinnati & Indianapolis)	Kentucky	July		15 x 22	60	4-4-0. Scrapped 1879
"	Sidney	July		15¼ x 22	72	4-4-0. Scrapped 1890's

*Niles is known to have built locomotives for the following railroads, but no specific data are available on the engines themselves: Evansville & Crawfordsville, Milwaukee & Watertown, Peru & Indianapolis, LaCrosse & Milwaukee, Racine & Mississippi, Dayton & Western Dayton & Michigan, and Kentucky Central.

Railroad	Name or number	Date delivered	Weight (tons)	Cylinders (inches)	Wheel diameter (inches)	Remarks
Cleveland, Columbus & Cincinnati	*Iowa*	April		16 x 22	48	4-6-0. Sold to U.S. Military Railroads 1862
"	*Wisconsin*	April		16 x 22	48	4-6-0
"	*Kansas*	May		16 x 22	48	4-6-0. Scrapped 1878
"	*Montana*	May		16 x 22	48	4-6-0. Scrapped 1878
"	*Minnesota*	May		16 x 22	48	4-6-0. Scrapped 1878
"	*Pennsyl- vania*	May		16 x 22	48	4-6-0. Scrapped 1878
"	*Michigan*	May		16 x 22	48	4-6-0. Scrapped 1878
Cincinnati, Hamilton & Dayton	*Chas. Hammond*	March	23½		48	4-4-0
Little Miami	*Jacob Strader*	June	23		66	4-4-0
"	*Alfred Kelly*	July	23		66	4-4-0
"	*J. H. Groesbeck*	Nov.	19		54	4-4-0
New Orleans, Opelousas & Great Western	*Natchitoches*	Nov.	18			4-4-0. Later Southern Pacific
"	*Texas*	Nov.	18			4-4-0. Later Southern Pacific
Ohio & Mississippi	*2*	Dec.				

NILES & COMPANY 1854

Railroad	Name or number	Date delivered	Weight (tons)	Cylinders (inches)	Wheel diameter (inches)	Remarks
Cincinnati, Hamilton & Dayton	*S. Gebhart*	May	22		66	4-4-0
Cincinnati, Wilmington & Zanesville	*Scioto*(?)	Jan.	25			4-4-0
"	*Zanesville*	Jan.	25		52	4-6-0
"	*Hocking*	April	25			4-6-0. Cost $9,889

156

Railroad	Name or number	Date delivered	Weight (tons)	Cylinders (inches)	Wheel diameter (inches)	Remarks
Indianapolis & Cincinnati	*Thomas Gaff*			16 x 22	46	4–6–0
"	*L. B. Lewis*			16 x 22	46	4–6–0
Memphis & Charleston	*Ohio*		29		51	Sold to U.S. Military Railroads 1865
Ohio & Mississippi	*3*	Feb.	28	16 x 20	60	4–4–0. Outside connected
"	*4*	Feb.	28	16 x 20	60	4–4–0. Outside connected. Scrapped 1889
"	*5*	Feb.	28	16 x 20	60	4–4–0. Outside connected
"	*6*	March	28	16 x 20	60	4–4–0. Outside connected
"	*7*	March	28	16 x 20	60	4–4–0. Outside connected
"	*8*	March	28	16 x 20	60	4–4–0. Outside connected
"	*9*	April	28	16 x 20	60	4–4–0. Outside connected
"	*10*	April	25½	15 x 20	60	4–4–0
"	*11*	April	25½	15 x 20	60	4–4–0
"	*12*	April	25½	15 x 20	60	4–4–0
"	*13*	May	26	15 x 20	60	4–4–0
"	*14*	May	26	15 x 20	60	4–4–0
"	*15*	July	26	15 x 20	60	4–4–0
"	*16*	July	26	15 x 20	60	4–4–0
"	*17*	Oct.	28	16 x 20	60	4–4–0
"	*18*	Oct.	28	15 x 20	60	4–4–0
"	*19*	Oct.	28	15 x 20	60	4–4–0
"	*20*	Nov.	28	16 x 20	60	4–4–0
"	*21*	Nov.	28	16 x 20	60	4–4–0
Ohio & Pennsylvania	*Pontiac*	Feb.				4–6–0
"	*Holmes*	Feb.				4–6–0
Springfield, Mt. Vernon & Pittsburgh	*Mercury*	Feb.	18	14½ x 20		Slab rail frame. Tested on Little Miami RR.

Railroad	Name or number	Date delivered	Weight (tons)	Cylinders (inches)	Wheel diameter (inches)	Remarks
NILES & COMPANY 1855						
Coal Run Improvement	*Defiance*	Aug.	30	16 x 20	44	0–8–0. Sold to Beaver Meadow RR.
"	*Champion*	Aug.	30	16 x 20	44	0–8–0. Sold to Beaver Meadow RR.
Indianapolis & Cincinnati	*Nat. Wright*					4–4–0
"	*T. A. Morris*					4–4–0
Little Miami	*R.W. McCoy*	March	19		54	4–4–0
Louisville & Nashville	*Hart County*	July	20	14 x 20	60	4–4–0. Out of service 1876
"	*Ben Spalding*	July	20	14 x 20	60	4–4–0. Out of service 1870
Marietta & Cincinnati	*Highland*	Nov.	25(?)	15 x 22	60	4–4–0
"	*Athens*	Nov.	25(?)	15 x 22	60	4–4–0
"	*Vinton*	Nov.	25(?)	15 x 22	60	4–4–0
New Orleans, Opelousas & Great Western	*Tenebonne*	Sept.	20			4–4–0
"	*Tiger*	Sept.	12			0–4–0
NILES & COMPANY 1856						
Little Miami & Columbus & Xenia	*Larz Anderson*	Nov.	22		66	4–4–0
Mississippi Central	*Richland*	Oct.	16	10 x 22	60	
NILES & COMPANY 1857						
Marietta & Cincinnati	*Chillicothe*	Jan.	26	16 x 22	66	
"	*Marietta*	Jan.	26	16 x 22	66	
"	*Baltimore*	Jan.	26	16 x 22	66	
"	*Philadelphia*	Jan.	26	16 x 22	66	
Mississippi Central	*D.B. Molloy*	May	20	14 x 22	60	
"	*James Brown*	May	20	14 x 22	60	

158

Railroad	Name or number	Date delivered	Weight (tons)	Cylinders (inches)	Wheel diameter (inches)	Remarks
New Orleans, Opelousas & Great Western	*Sabine*	June	20	12 x 22	4–4–0.	Scrapped 1942
"	*New Orleans*	June	20		4–4–0	

NILES & COMPANY 1858

Railroad	Name or number	Date delivered	Weight (tons)	Cylinders (inches)	Wheel diameter (inches)	Remarks
Chicago & Northwestern	*Elkhorn*				4–4–0. May have been built earlier for Galena & Chicago Union RR.	
"	*Richmond*				4–4–0. May have been built earlier for Galena & Chicago Union RR.	

GEORGE ESCOL SELLERS 1851–1852

Railroad	Name or number	Date delivered	Weight (tons)	Cylinders (inches)	Wheel diameter (inches)	Remarks
Panama		Aug. 1851	12		42	
"	*Isthmus*	Feb. 1852	12		42	
"		March 1852	12		42	

COVINGTON LOCOMOTIVE WORKS 1853–1857

Railroad	Name or number	Date delivered	Weight (tons)	Cylinders (inches)	Wheel diameter (inches)	Remarks
Covington & Lexington	*Covington*	May 1854	24			
"	*Cynthiana*	June 1854	24			
"	*Paris*	July 1854	24			
"	*Lexington*	Aug. 1854	24			

Railroad	Name or number	Date delivered	Weight (tons)	Cylinders (inches)	Wheel diameter (inches)	Remarks
Covington & Lexington	M.M. Benton	June 1857	22			
"	Sam J. Walker	Sept. 1857	22			
U.S. Military	General Walker	?		16 x 24	54	Probably bought secondhand
"	General Allen	?	28	15 x 22	54	Probably bought secondhand

160

Bibliography

M. W. Baldwin papers: account and letter books [c. 1835–1866]. MSS., Historical Society of Pennsylvania, Philadelphia.

Cist, Charles. *Sketches and statistics of Cincinnati in 1859.* Cincinnati: Charles Cist, 1859.

Greve, Charles T. *Centennial history of Cincinnati.* Vol. 1. Cincinnati, 1904.

In Memoriam, Cincinnati 1881, containing proceedings of the Memorial Association, eulogies at Music Hall, and biographical sketches of many distinguished citizens of Cincinnati. Vol. 1 [no further volumes were published]. Cincinnati: A. E. Jones, 1881.

Marvin, Walter R. Columbus and the railroads of central Ohio before the Civil War. Doctoral dissertation, Ohio State University, 1953.

Meyer, B. H. *History of transportation in the United States before 1860.* Washington, D.C., 1917.

Moore, Robert. *Autobiographical outlines of a long life.* Cincinnati, 1887.

Otis, F. N. *Illustrated history of the Panama Railroad.* New York, 1861.

Panama Railroad: minute books of directors and executive committee [1848–c. 1855]. MSS., U.S. National Archives, Washington, D.C.

Patent Office papers. U.S. National Archives, Washington, D.C.

Peale-Sellers papers: letters and accounts of George Escol Sellers and Coleman Sellers, Jr. MSS., American Philosophical Society, Philadelphia, Pennsylvania.

Edwin Price reminiscences [1829–1901]. MS., U.S. National Museum, Washington, D.C.

Report on affairs of Southern Railroads. 39th Congress, 2d session. House report no. 34, 1867.

Sellers, George E. *Improvements in locomotive engines and railways.* Cincinnati: Wright, Fisher & Co., 1849.

Sinclair, Angus. *Development of the locomotive engine.* New York: D. Van Nostrand Co., 1907.

Smith, William P. *Great railway celebration of 1857.* New York, 1858.

Watkins, J. Elfreth. History of the Pennsylvania Railroad. Unpublished proof sheets [1896]. U.S. National Museum, Washington, D.C.

Of the numerous periodicals consulted in the preparation of this study (as specific citations indicate), the following were among the most helpful: *American Machinist, American Railroad Journal, American Railway Times, Engineering* (London), *Farmer and Mechanic, Journal of the Franklin Institute, Scientific American, Scientific Artisan, Railroad Advocate,* and *Railroad Record.*

Index

163

164

165

valve gears, 128, 141; radial, 100, 102; shifting-link, 37; stationary-link, 37, 39; suspended-link, 37, 128; Walschaert, 100
valve stem, 39
Van Loan. *See* Van Loon.
Van Loon, William, 12, 22, 31
Vigneols, Charles B., 47
Vulcan Foundry (Tayleur & Company, Warrington, England), 100

wagon-top boilers, 35, 36, 40, 107, 109
Walschaert valve gear, 100

Western Star, 7
Wm. McCammon, 126
W. S. Watterson, 35
wheels: adhesion, 47; driving, 53; gripper, 53; wrought-iron driving, 79
Whetstone, John L., 23, 58, 61, 62, 76, 93, 100, 105, 135, 138, 139
Wilmarth, Seth, 105
Winans, Ross, 36, 105
Wolff, Daniel, 121, 122
Woodward, Jabez M., 62

Yeatman and Shield, 22
Yuma, 44

U.S. GOVERNMENT PRINTING OFFICE:1965

A Moore & Richardson Locomotive of about
1855. Road and number unknown. From a
colored lithograph. (*Courtesy John W. Merten.*)